Phytotechnology with Biomass Production

Phytotechnology with Biomass Production

Sustainable Management of Contaminated Sites

Edited by
Larry E. Erickson and Valentina Pidlisnyuk

CRC Press
Taylor & Francis Group
Boca Raton London New York

CRC Press is an imprint of the
Taylor & Francis Group, an **informa** business

First edition published 2021
by CRC Press
6000 Broken Sound Parkway NW, Suite 300, Boca Raton, FL 33487-2742

and by CRC Press
2 Park Square, Milton Park, Abingdon, Oxon, OX14 4RN

Library of Congress Cataloging-in-Publication Data
Names: Erickson, L. E. (Larry Eugene), 1938- editor. |
Pidlisnyuk, Valentina V., editor.
Title: Phytotechnology with biomass production : sustainable management of
contaminated sites / edited by Larry E. Erickson and Valentina Pidlisnyuk.
Description: First edition. | Boca Raton : CRC Press, 2021. |
Includes bibliographical references and index.
Identifiers: LCCN 2021021263 (print) | LCCN 2021021264 (ebook) |
ISBN 9780367522803 (hardback) | ISBN 9781003082613 (ebook)
Subjects: LCSH: Phytoremediation. | Soil remediation. | Energy crops.
Classification: LCC TD878.48 .P49 2021 (print) | LCC TD878.48 (ebook) |
DDC 628/.74—dc23
LC record available at https://lccn.loc.gov/2021021263
LC ebook record available at https://lccn.loc.gov/2021021264

ISBN: 978-0-367-52280-3 (hbk)
ISBN: 978-0-367-53620-6 (pbk)
ISBN: 978-1-003-08261-3 (ebk)

Typeset in Palatino
by codeMantra

Contents

Preface

Land stewardship and management are important topics in the Earth Charter and United Nations Sustainable Development Goals. The processes to improve soil quality and land use have global importance. Research, development, and practices to remediate land are advancing with new approaches in many countries. Great progress has been made in the science and engineering related to phytotechnologies for contaminated soil during the years from 1990 to the present.

In February, 1989, the Great Plains/Rocky Mountain Hazardous Substance Research Center was established with headquarters at Kansas State University and funding from the U.S. Environmental Protection Agency. Research on the beneficial effects of vegetation in contaminated soil became an important thrust of the center at several of the participating universities. One of the early projects involving phytoremediation with biomass production investigated the impact of trees planted at the edge of the field to reduce the amount of nitrogen that goes into the stream when storm water flows off of the field. From this early start, much research to develop phytotechnologies to address contamination problems in soil and ground water has been conducted, and the results have been reported. The efforts to advance the science of phytotechnologies have been helped by the International Phytotechnology Society and by a number of international events including NATO Science for Peace and Security Program Advanced Research Workshop in Ukraine in 2007.

Economic analysis has shown that phytotechnologies are the most cost-effective options because soil quality is improved and costs associated with applications are low. The five hazardous substance research centers were established by U.S. Environmental Protection Agency to address contamination problems where risk needed to be reduced. Land improvement is an important goal of the Earth Charter, U.N. Sustainable Development Goals, and U.N. Food and Agriculture Organization. There are many locations where improvements in soil quality are beneficial and application of phytotechnology with biomass production is not driven by risk reduction, but rather by the goal of improving soil quality and site productivity. With the growth in population and the need to reduce the concentration of greenhouse gases in the atmosphere, all land needs to be used effectively. There are presently many locations where phytotechnologies can be applied to improve soil quality and make better use of the site.

In 2004 Valentina Pidlisnyuk visited Kansas State University as part of her Fulbright scholar program. After this visit cooperative efforts related to phytotechnologies with biomass production were established and carried out. The preparation of manuscripts for this book is one of the activities of the NATO Science for Peace and Security Program Multiyear Research Project

G4687 entitled Phytotechnology for Cleaning Contaminated Military Sites. Many of the results of the NATO project have been published in journals with citations and references included in this book. The contents of the book include results from research completed as part of the NATO project but also results from other studies. From 1990 to the present, there has been a great progress related to sites with metals and sites with organic contaminants. Military sites and lands contaminated by mining, pesticides, petroleum, and chlorinated solvents are included.

One of the reasons to make use of phytotechnology with biomass production at contaminated sites is to restore the site to a useful state where biomass is being produced, soil quality improvement is happening, and organic carbon is being sequestered in the soil. The process of establishment of vegetation after an industrial site or mining site has been closed can benefit from the application of soil amendments and efforts to modify the physical properties of the soil. The health of the ecosystem can be improved through better understanding of the biological populations and their contributions to soil health.

Miscanthus is a large perennial and rhizomatous grass imported to Europe from East Asia as an ornamental plant in the 1930s. In 1990 it started to be investigated as an energy crop, and since 2000 there has been research and development on applications of its biomass as a raw material in different industries. While there are a variety of *Miscanthus* species, most of the research has been with *Miscanthus* ×*giganteus* (*M.*×*giganteus*), which is a hybrid of *Miscanthus sinensis* and *Miscanthus sacchiflorus*. *Miscanthus* shows the highest harvest among second-generation crops; it can reach a height of 3 m; the crop has a C4 photosynthetic pathway, an excellent environmental profile to increase soil carbon, nutrients, and biodiversity. *Miscanthus* needs to be planted only once; it has the potential to provide annual harvests for 20 years, because it is a perennial, nutrient runoff and leaching are small. The yield of a fully established plantation is often between 10 and 20 t ha^{-1} on a dry matter basis with values as large as 25–30 t ha^{-1} depending on local conditions and agricultural practices. The energy content is comparable to wood with values around 17 MJ t^{-1} of dry biomass.

Since 2000 there has been an increased effort to grow *M.* ×*giganteus* on contaminated soils at postindustrial and former military sites and on marginal and abandoned lands. The biomass yield is often lower, and this has stimulated phytomanagement research on improving soil quality and biomass production. Generally two main approaches are under consideration: adding different amendments to improve soil quality and actions to affect the rhizomes and plants such as adding plant growth regulators. Soil amendments include fertilizers, biosolids, biochar, manure, sludge, compost, and activated carbon. The addition of microorganisms such as endophytes and plant growth-promoting bacteria and chemicals such as plant growth regulators are beneficial for *Miscanthus*.

Miscanthus biomass can be used for different purposes; it contains cellulose, hemicellulose, and lignin. Cellulose has important applications in bio-based materials including paper. Hemicellulose protects cellulose from enzymatic hydrolysis, and lignin is primarily responsible for recalcitrance to chemical and enzymatic degradation. *Miscanthus* removes carbon from the atmosphere as it grows; this carbon can be incorporated into building materials and stored as part of a building for many years. The importance of making products from renewable biomass has received significant attention recently.

The book includes chapters on phytotechnologies for inorganic contaminants where phytostabilization is important and organic contaminants where biodegradation is desired. Phytomining is included because of its growing importance and recent progress. The science and technology on establishing vegetation at contaminated sites and the benefits of adding soil amendments and plant growth regulators are discussed as well. Phytotechnologies improve soil health and ecosystem services as well as produce valuable biomass. There is a chapter on plant feeding insects, nematodes, and other topics associated with *Miscanthus* plant health. Two chapters address products of *Miscanthus*: one is on bioenergy as an alternative of growing importance while the other is on biomass as a material for composites, building materials, and biodegradable and renewable products to replace plastics from petroleum.

This book is for faculty, students, research scientists, environmental and agricultural professionals, gardeners, farmers, landowners, and government officials. An important goal for this book is to have value for all who are working on phytotechnology projects to reduce risk and/or improve soil quality at contaminated sites. It will also be interesting for those who are working to improve soil quality on marginal and abandoned lands. The book will be very beneficial for those who are new to the topics and want to learn to apply phytotechnologies and biomass production with its further converting to energy and bioproducts.

Larry E. Erickson
Valentina Pidlisnyuk

Acknowledgments

The NATO-supported Multiyear Research Project of the Science for Peace and Security Program entitled Phytotechnology for Cleaning Contaminated Military Sites has supported research in the Czech Republic, Ukraine, Kazakhstan, Poland, Croatia, and the United States with matching funds provided by the participating universities and research organizations, i.e., Jan Evangelista Purkyne University, Kansas State University, National University of Life and the Environmental Science, National University "Lvivska Polytechnilka", Institute of Plant Biology and Biotechnology. Thank you for the financial support and the related efforts to provide research laboratory space and equipment, and administrative support. Many individuals have worked on research projects, helped with supporting tasks, and assisted with efforts to make newly published literature available, first of all young researchers and PhD students from NATO teams: Artem Medkow, Bulat Kenessov, Ethan Duong, Hana Malinská, Iwona Gruss, Kamilya Abit, Kumar Pranaw, Lyudmila Kava, Maria Ovruch, Martyn Sozanskyi, Svitlana Yaschuk, Vitaliy Stadnik, Volodimir Kvak, and Zafer Alasmary who have been involved in the Lab and Field NATO project research in Ukraine, Kazakhstan, and the United States, analysis and evaluation of the numerous data. KSU students have worked on design projects related to potential uses of *Miscanthus* in building applications and energy. Thank you to all who have helped.

Many authors have helped with chapters in order to include important content. Thank you to all authors, to Danita Deters and Aigerim Mamirova for helping with the manuscripts, figures, and tables.

We thank Zafer Alasmary, Marek Bury, Lawrence Davis, John Dolman, and Mark Janzen for permission to use their photos of plants.

We thank Fort Riley and the U.S. Department of Army for use of military land for a field research site, Dolyna regional council and personally Volodymyr Garazd[†], Mayor of Dolyna, Ukraine for use of the postmilitary site, Tokarivka regional farmers, Ukraine for assistance with the field research and active participation in the project's events. We specifically thank Oleksandr Mazurchak, First Vice-President of "Mayor's Club", Ukraine, for the constant professional support during the NATO-supported Multiyear Research Project's life.

[†] Mr. Garazd passed away on January 6, 2021.

Editors

Larry E. Erickson has been associated with chemical engineering at Kansas State University, USA, since 1957 as a student and since earning a PhD in 1964 as a faculty member. In 1985, he helped to establish a research program at K-State to address hazardous substance issues. From 1989 to 2003 he directed the Great Plains/Rocky Mountain Hazardous Substance Research Center, with financial support from the U.S. Environmental Protection Agency and other sources. This consortium of universities began conducting research with vegetation and helped to develop phytotechnologies to address contaminated soil problems. He serves on the editorial board of the *International Journal of Phytotechnology* and the journal *Environmental Progress and Sustainable Energy*. He is currently an emeritus professor of chemical engineering at K-State. He has helped to provide leadership for the Center for Hazardous Substance Research at K-State since 1989 and has been part of the leadership team at K-State for the NATO project that Valentina Pidlisnyuk leads.

Valentina Pidlisnyuk, DrSc, serves as professor at the Department of Environmental Chemistry and Technology, Jan Evangelista Purkyne University in Usti nad Labem, the Czech Republic. Her research interests include sustainability, phytotechnologies, value chain of biomass and bio-products, and environmental policy and management. She teaches PhD and master's courses in phytoremediation and Erasmus graduate courses in sustainable management of contaminated sites, fundamentals of sustainability, and global environmental change. She serves as a topic editor at the Plant and the *Journal of Elementology* and is a member of the editorial board of the *Central and Eastern European Journal of Management and Economics*. Prof. Pidlisnyuk has worked with more than 50 graduate students and advised four PhD students who successfully defended their theses. Currently she is advising three PhD students.

Prof. Pidlisnyuk received a professorship in Environmental Sciences from the Ukrainian Ministry of Education and Science, was confirmed by the Ministry of Education and Science of Slovakia, earned a doctorate in colloidal chemistry at the Institute of Colloidal and Water Chemistry, National Academy of Science of Ukraine, and a master's degree in chemistry at the National State University, Ukraine, confirmed by Carl University in Prague, Czech Republic. She accomplished a Fulbright Research Program at the University of Georgia, USA, and an Environmental Management Program at the Japan International Cooperation Agency, Japan.

Contributors

Nikola Bilandžija
Faculty of Agriculture
University of Zagreb
Zagreb, Croatia
0000-0001-9513-958X

Jan Černý
Department of Environmental
Chemistry and Technology
Jan Evangelista Purkyně University
Ústí nad Labem, Czech Republic
0000-0001-5823-4537

Lawrence C. Davis
Department of Biochemistry and
Molecular Biophysics
Kansas State University
Manhattan, Kansas
0000-0002-9044-0282

Larry E. Erickson
Tim Taylor Department of Chemical
Engineering
Kansas State University
Manhattan, Kansas
0000-0001-7012-4437

Hermann Heilmeier
Institute of Biosciences
Technische Universität
Bergakademie Freiberg
Freiberg, Germany
0000-0001-5935-0396

Ganga M. Hettiarachchi
Department of Agronomy
Kansas State University
Manhattan, Kansas
0000-0002-6669-2885

Aigerim Mamirova
Department of Biotechnology
Al-Farabi Kazakh National
University
Almaty, Kazakhstan
0000-0002-4274-5081

Diana Nebeská
Department of Environmental
Chemistry and Technology
Jan Evangelista Purkyně University
Ústí nad Labem, Czech Republic
0000-0002-4388-5297

Asil Nurzhanova
Laboratory of Plant Physiology and
Biochemistry
Institute of Plant Biology and
Biotechnology
Almaty, Kazakhstan
0000-0003-4811-0164

Valentina Pidlisnyuk
Department of Environmental
Chemistry and Technology
Jan Evangelista Purkyně University
Ústí nad Labem, Czech Republic
0000-0002-1489-897X

Melissa Prelac
Paying Agency for Agriculture
Fisheries and Rural Development
Zagreb, Croatia

Kraig Roozeboom
Department of Agronomy
Kansas State University
Manhattan, Kansas
0000-0003-1225-5177

John Schlup
Tim Taylor Department of Chemical
    Engineering
Kansas State University
Manhattan, Kansas

Pavlo Shapoval
Department of Physical, Analytical
    and General Chemistry
National University Lvivska
    Polytechnika
Lviv, Ukraine

Andrzej Skwiercz
Department of Plant Protection
Research Institute of Horticulture in
    Skierniewice
Skierniewice, Poland

Tatyana Stefanovska
Department of Entomology
National University of Life and
    Environmental Sciences of
    Ukraine
Kyiv, Ukraine

Josef Trögl
Department of Environmental
    Chemistry and Technology
Jan Evangelista Purkyně University
Ústí nad Labem, Czech Republic

Donghai Wang
Department of Biological and
    Agricultural Engineering
Kansas State University
Manhattan, Kansas

Barbara Zeeb
Department of Chemistry and
    Chemical Engineering
Royal Military College of Canada
Kingston, Ontario, Canada

Zeljka Zgorelec
Department of General Agronomy
University of Zagreb
Zagreb, Croatia

Jikai Zhao
Department of Biological and
    Agricultural Engineering
Kansas State University
Manhattan, Kansas
0000-0002-0119-8640

1

Introduction

Larry E. Erickson and Valentina Pidlisnyuk

Abstract

Land management is an important sustainability challenge in many locations because of contamination and/or degradation. There are many aspects related to reducing risk associated with contaminants in soil and improving soil quality. The science and engineering of phytotechnologies with biomass production have been advanced in many countries with much of the research being published after 1985. In most phytotechnology applications at contaminated sites, improving soil quality is a priority, and productive use of the land is one of the goals. Remediation of contaminated soil has great value for society and being able to produce a biomass product to improve the economics is very beneficial in countries where resources to address environmental challenges are limited. Miscanthus is a very valuable biomass product and an effective plant for many applications because of its properties. Since partial funding has been provided by NATO, an effort has been≈made to address topics that are important to NATO countries in this work.

CONTENTS

1.1 Soil Quality

Soil quality is a high priority because of food and other useful products that can be produced through agriculture and forestry. Each plot of land has features that can be nurtured in order to improve soil quality and crop productivity. Stewardship of soil and the restoration of contaminated sites are important in many countries.

The soil scientist characterizes soil in terms of physical properties such as particle size distribution, chemical properties such as pH and percent nitrogen (N), and biological properties such as number of microorganisms per gram of soil. Soil quality depends on having a desirable amount of many different substances in soil.

1.1.1 Soil Contamination

This book has a focus on phytotechnologies for contaminated sites. The NATO funding for multiyear research project of the Science for Peace and Security Program (SPS MYP) #G4687 entitled Phytotechnology for Cleaning Contaminated Military Sites has a goal to improve the economics of phytoremediation by producing plant biomass that has economic value. There are many locations in the world which are contaminated with substances in the soil that reduce the value of the products that are harvested because yield is reduced or the products are of lower quality. In some cases, health and safety issues may prevent the land from being used for the production of food and feed crops. Risk reduction is one of the important issues in many locations where contaminants are present.

There is value in approaching the topics in this book using a sustainability approach in which social value, environmental quality, and economic benefits are all considered to be important. Each site with contamination has the potential to be improved such that it can be used productively for the benefit of society. Many investigators have reported on their efforts to develop methods and approaches that advance the science and engineering associated with phytotechnologies. One of the goals of this book is to collect and write about some of these developments.

The research site at Fort Riley, Kansas, where Miscanthus has been established has lead in soil. Because of its use for military purposes, lead in soil is a contaminant of concern at a number of military sites. At a research site at Dolyna, Ukraine, there are other inorganic contaminants in the soil.

Petroleum contaminated soil is common on military lands and at sites where vehicles are refueled. Petroleum refineries often have one or more contaminated areas. Pipeline spills are commonly found in many countries, and the spill may be due to military operations designed to disrupt supplies. Pesticides are often contaminants of concern on military lands and at other locations.

There are many locations where mining operations have ended with land that needs to be restored to better quality. Minerals, coal, lime, and other products are mined to obtain useful raw materials, but there are other solid residues that remain at the site. These mine tailings are often an important challenge in efforts to improve soil quality.

Salt is a contaminant of concern when it is present in soil at a high concentration. High salt concentrations may be from oil and gas production or irrigation water that has too much salt.

Coal ash is a residue from coal combustion for electricity or to fire a boiler at an industrial site. There are also residues from other industrial processes such as steel mills.

1.1.2 Types of Contaminants

There are two important types of contaminants in soil: inorganic and organic substances. Organic contaminants such as gasoline can be biodegraded by biological processes to carbon dioxide and water. Inorganic contaminants such as lead, other trace elements such as zinc, and salts do not biodegrade. They are managed by other processes such as phytostabilization. Both types of contaminants are important and are considered in this book.

1.2 Phytotechnology with Biomass Production

One of the core topics in this book is how to restore soils that are contaminated. There are many ways to make use of the beneficial effects of vegetation in contaminated soil. The concept of using plants to both improve soil quality and produce a product that has commercial value has been investigated by a number of research teams and used in practice (Alexopoulou 2018; McCutcheon and Schnoor 2003; Nsanganwimana et al. 2014; Pidlisnyuk et al. 2014). This is one of the core topics of this book, which will be described further in other chapters. While the NATO project research has been with Miscanthus, there are many other plants that have the potential to be used commercially to both improve soil quality and harvest a product that can be marketed and sold.

1.3 Miscanthus

The focus of this book includes growing Miscanthus at contaminated sites in order to have a product to harvest and sell. Miscanthus has been the subject of considerable research because of the large amount of biomass that can be produced (Alexopoulou 2018; Lewandowski et al. 2000; Jones 2020). Miscanthus crops may exceed 3 m in height and annual yields may be more than 25 Mg ha^{-1} (Jones 2020). Miscanthus genetics and the development of hybrids that have desirable features are important. Establishing Miscanthus plants and other vegetation in contaminated soil is the subject of a separate chapter because of its importance. Soil fauna have great value in improving soil quality when Miscanthus is grown in contaminated soil.

The markets for Miscanthus include its use as a biofuel and as a bio-based product in buildings, paper industry, vehicles, or other applications. The economics associated with the marketing of Miscanthus vary with location and with the efforts in various countries to have policies to reduce greenhouse gas emissions.

1.4 Case Studies

This book includes several recent examples of efforts to produce Miscanthus while also improving soil quality at field sites in several countries. At each field site, there are applications of the science and technology described in the various chapters. The details of many of the case studies are presented in other publications which are included in the references associated with each case study.

References

Alexopoulou, E. Ed. 2018. *Perennial grasses for bioenergy and bioproducts. Production, uses, sustainability and markets for giant reed, miscanthus, switchgrass, reed canary grass and bamboo*. London: Elsevier, Academic Press, 306 p. https://doi.org/10.1016/C2016-0-03729-4.

Jones, M.B. 2020. *Miscanthus for bioenergy production. Crop production, utilization and climate change mitigation*. New York: Routledge, 120 p. ISBN 9781138091245.

Lewandowski, I., Clifton-Brown, J.C., Scurlock, J.M.O., Huisman, W. 2000. Miscanthus: European experience with a novel energy crop. *Biomass and Bioenergy*, 19: 209–227. https://doi.org/10.1016/S0961-9534(00)00032-5.

McCutcheon, S.C. and J.L. Schnoor. Eds. 2003. *Phytoremediation: Transformation and control of contaminants*. Hoboken, NJ: Wiley, 987 p. ISBN 0-471-39435-1.

Nsanganwimana, F., Pourrut, B., Mench, M., Douay, F. 2014. Suitability of Miscanthus species for managing inorganic and organic contaminated land and restoring ecosystem services. A review. *Journal of Environmental Management*, 143: 123–134. https://doi.org/10.1016/j.jenvman.2014.04.027.

Pidlisnyuk, V., Stefanovska, T., Lewis, E.E., Erickson, L.E., Davis, L.C. 2014. Miscanthus as a productive biofuel crop for phytoremediation. *Critical Reviews in Plant Science*, 33(1): 1–19. https://doi.org/10.1080/07352689.2014.847616.

2

Phytotechnologies for Site Remediation

Valentina Pidlisnyuk, Ganga M. Hettiarachchi,
Zeljka Zgorelec, Melissa Prelac, Nikola Bilandžija,
Lawrence C. Davis, and Larry E. Erickson

Abstract

Phytotechnologies for inorganic contaminants include phytoextraction, phytostabilization, phytotransformation, and phytohydraulics. Soil amendments may be added to increase contaminant solubility when phytoextraction is implemented. For phytostabilization soil amendments may be added to reduce contaminant availability, such as transformation to a less soluble compound. Phytotransformation is the process of changing the contaminant to another form to reduce risk of movement or toxicity. Phytohydraulics may be applied with phytostabilization when the design includes evapotranspiration to reduce transport of the contaminants away from the point of contamination. Plants used for phytostabilization should be able to grow well in the contaminated soil, produce a product of value and commercial interest, and evapotranspire sufficient water to achieve containment of the contaminants. The uptake and translocation of the contaminants to aboveground biomass should be small enough to allow the plant biomass to be used for a commercial purpose. Miscanthus is among the most promising energy crops for phytoremediation: it grows well in contaminated soil, evaptranspires large quantities of water, and produces high-quality cellulose. The use of soil amendments can help to minimize contaminant uptake and improving soil quality is an important issue. Several other energy crops that have good potential for phytostabilization application are introduced in this chapter as well.

CONTENTS

2.1 Introduction

Phytoremediation technology was established as an environment-friendly concept to restore polluted sites before 1990, and a year later the term was used. Generally, phytoremediation technology can be divided into six subtypes depending on the contaminant origin and the mechanisms of restoration (USEPA, 1998). In order to remediate sites polluted by contaminants of inorganic origin (mainly trace elements) the following processes are proposed (USEPA, 1998):

1. phytoextraction – remediation mechanism is based on the uptake of contaminant by roots and its transmigration to the aboveground tissues (leaves, stems, branches);

2. rhizofiltration – remediation mechanism is represented by the accumulation of contaminants in roots;

3. phytostabilization – remediation mechanism is based on the contaminant immobilization in soil by plant root exudates.

In case of remediation sites polluted by contaminants of organic origin the following processes are in the focus (USEPA, 1998):

4. phytodegradation – remediation mechanism is based on absorbing the contaminants by roots and converting them by plant enzymatic activity to safe compounds;

5. rhizodegradation – remediation mechanism is based on the provision of favorable environmental conditions for microorganisms to

be able to degrade the contaminants in the rhizosphere, plant roots release organic compounds (nutrients, enzymes, organic acids, etc.) to reach favorable conditions;

6. phytovolatilization – remediation mechanism is based on absorbing the contaminants by roots and releasing them to atmosphere.

The term phytotechnologies, which replaces the earlier term phytoremediation, is also known as green-remediation, and in general, means using plants to degrade, extract, contain, transform into less harmful forms, or immobilize contaminants in soil, water, or air with inorganic or organic compounds. Phytotechnology mechanisms and technological effects are summarized in Table 2.1. Some phytotechnologies have gained public acceptance compared to other remediation techniques and multiple related terms, such as phytomanagement, have been introduced. As far as phytomanagement is concerned, this concept is newer and covers economic benefits (Robinson et al., 2009). In addition to using plants to reduce the risks posed by soil contamination, the concept

TABLE 2.1

Phytotechnology Mechanisms and Effect of Technology

Effect of Technology	Phytotechnology Mechanism	Definition
Reduce contaminant concentration (extraction, degradation)	Phytoextraction/ Phytomining/ Phytoaccumulation	The removal of inorganic contaminants from the soil through plant uptake, and subsequent harvest and removal of biomass. Phytoextraction, phytomining, or phytoaccumulation are typically used to remove metals from the soil (e.g., As extraction by *Pteris vittata* (brake fern) (Ma et al., 2001); Ni by Alyssum species (Li et al., 2003).
	Phytodegradation/ Phytotransformation	The breakdown of contaminants by the metabolic processes in a plant. Also includes the breakdown of contaminants in the soil by enzymes or other products produced by the plant. Primarily used for organic contaminants.
	Rhizofiltration/ Rhizodegradation	The breakdown or degradation of organic contaminants in the soil. The contaminants are either adsorbed onto the root surface or are absorbed by the plant roots. Due to enhanced microbial activity in the rhizosphere (the zone of soil influenced by the roots), the contaminants are broken down. This process can be enhanced by fertilization.

(Continued)

TABLE 2.1 (*Continued*)

Phytotechnology Mechanisms and Effect of Technology

Effect of Technology	Phytotechnology Mechanism	Definition
	Phytovolatilization	The uptake of contaminants by plants and release them into the atmosphere as they transpire water (direct phytovolatilization). Contaminant is removed from the soil and may be degraded as it moves through the plant's vascular system before final removal from the system. This can be used for both organic (e.g., volatile organic compounds) and inorganic contaminants (e.g., Se, and Hg). Additionally, contaminants can be volatilized from soil due to plant root activities (indirect phytovolatilization) (Limmer & Burken, 2016)
Reduce contaminant bioavailability without reducing total concentrations (immobilization)	Phytosequestration/ Phytostabilization	This process sequesters, or reduces, contaminant bioavailability through precipitation or immobilization of contaminants in the soil, on the root surface, or within the root tissues (Laperche et al., 1996).
	Phytotransformation (inorganics)	The transformation of contaminants by the metabolic processes in a plant. It also includes the transformation of contaminants in the soil by enzymes or other products produced by the plant. This can be used for nutrients and other inorganic contaminants.
Reduce contaminant movement (containment)	Phytostabilization (Phytorestoration, In place inactivation)	Plants are used to stabilize contaminated soils or sediments, thus protecting them from transport by wind or water erosion. The main function is to contain the contaminated material. However, this is usually combined with adding soil amendment to reduce contaminant movement in soil (e.g., phytostabilzation of Pb contaminated military site soil using Miscanthus in combination with P amendments (Alasmary et al., 2020)).
	Phytohydraulics	This process is used to limit the movement of contaminants with water. Plants are used to increase evapotranspiration, thereby controlling soil water and contaminant movement. This mechanism contains the contaminant by modifying site hydrology to reduce the vertical or horizontal movement of water in the soil (Narayanan et al., 1999).

includes converting the obtained biomass into useful products (Evangelou et al., 2015; Robinson et al., 2009); in other words, "phytomanagement is a combination of phytoremediation and sustainable site management with economic return" (Conesa et al., 2012; Pandey & Bajpai, 2019).

The goal of this section is to provide an overview of various plant-based techniques for remediation of contaminated soils and to introduce research and applications of plant cultivation in contaminated areas to obtain biomass, with emphasis on Miscanthus as the phytoagent (Pidlisnyuk et al., 2014).

2.2 Phytotechnologies

The two most commonly used phytotechnologies for inorganic contaminants are phytoextraction and phytostabilization. In the phytoextraction process the plant's ability to accumulate trace elements is important. Based on the relative uptake and bioaccumulation potential, plants can be grouped into three categories: excluders, indicators, and accumulators (Adriano, 2001; Hunt et al., 2014) (see Figure 2.1).

The ability of plants to accumulate trace elements from the soil can be estimated by the enrichment coefficient (EC) or the bioconcentration factor, which are expressed as the ratio of defined trace element concentrations in the plant material (mg kg^{-1} of dry matter) and in the soil (mg kg^{-1} of dry soil). In addition, translocation factor (TLF) value reflects the levels of plants' phytoextraction potential accounting as a ratio of the contaminants' concentrations in the aboveground biomass to their concentration in roots

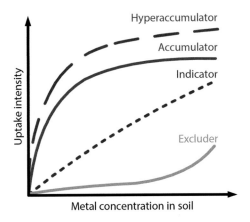

FIGURE 2.1
Three groups of plant categories based on their uptake behavior of trace elements: excluders, indicators, and accumulators. (Modified from Hunt et al., 2014.)

(Baker, 1981; Mamirova et al., 2020). Malayeri et al. (2008) and Zgorelec (2009) grouped plant species according to their ECs uptake capacities related to trace elements and sensitivity to trace element contamination as follows:

high-accumulator plants EC 1–10

medium-accumulator plants EC 0.1–1

low-accumulator plants EC 0.01–0.1

nonaccumulator plants EC < 0.01

The sensitivity or tolerance of plants to excess trace elements depends on plant species and their genotypes. Even among crops, sensitivity varies widely (Adriano, 2001). Excluders are tolerant to high-trace elements concentrations and their tolerance is achieved by preventing the absorption and translocation of toxic elements to aboveground biomass. However, the observed concentrations in the root are lower than those in the soil. These species have low potential for metal extraction, and EC value is less than one (Lasat, 1999). They are insensitive to trace elements over a wide range of concentrations, and a common example is various grass species. Indicators are plant species sensitive to trace elements and consequently certain symptoms can be manifested. Such plants are not tolerant to high concentrations of trace elements, and examples are grain and cereal crops (Adriano, 2001). Accumulators are plants which accumulate trace elements in various concentrations, however, in concentrations which are still lower than in case of hyperaccumulators. Hyperaccumulation is a specific characteristic inherent in certain species. The majority of hyperaccumulators are endemic plants which grow in soils rich in trace elements and behave like strict metallophytes (Baker & Brooks, 1989), where certain facultative metallophytes can survive in soils poor in the trace elements (Rascio & Navari-Izzo, 2011). These plants have developed special mechanisms for element uptake and tolerance to high concentrations. The mechanisms are genetically conditioned because the plant tissues do not manifest toxicity symptoms typical in plants due to high element concentrations. For successful hyperaccumulation the concentrations of the trace elements in the above-ground biomass have to be extremely higher compared to their content in the soil (Pulford & Watson, 2003). The negative aspect of hyperaccumulators is low biomass yield.

2.3 Phytostabilization of Arable Land Contaminated with Trace Elements

Phytostabilization uses plants to reduce contaminated soil material movement via water or wind. The approach is commonly used in combination with soil amendments to reduce contaminant bioavailability and to assist plant growth.

The term "bioavailability" is defined as the fraction of an element in soil which is available for absorption by humans, animals, or plants. In the processes of *in situ* inactivation or chemical stabilization the appropriate soil amendments are used for immobilization of contaminants in soil (Attanayake et al., 2014; Berti & Cunningham, 1997; Brown et al., 2003; Defoe et al., 2014; Hettiarachchi et al., 2000). Contaminant immobilization is achieved through enhanced sorption in soil, absorption and accumulation in roots, adsorption to roots or precipitation within the root zone, and physical stabilization of soils (USEPA, 2000). Phytostabilization assisted with soil amendments can also support re-establishment of vegetation at contaminated sites lacking native grasses because of high concentrations of phytotoxic trace elements or poor soil characteristics, i.e., low pH and poor agrichemical and physical characteristics (Gudichuttu, 2014; Solís-Dominguez et al., 2012; Tordoff et al., 2000; Wijesekara et al., 2016). Phytostabilization is recommended when other remediation approaches are not feasible due to extended contaminated area. It is also useful in case of limited funds for another remediation technique. The advantages of phytostabilization are low cost, simple implementation, and aesthetic aspects (Berti & Cunningham, 2000). The disadvantages are as follows: needs a long period (usually 30–40 years) for restoration of the certain area, problems connected with disposal of contaminated biomass, limited development of root system, seasonal and climate dependence.

There are two types of plants suitable for a phytostabilization process. One group is formed by plants tolerant to high concentrations of contaminants, i.e., trace element excluders. Another group is represented by species with highly developed root systems that can immobilize contaminated substances through uptake, precipitation, or reduction. In this case rather often relevant contaminants are concentrated in the roots and only a small portion of contaminants can move to the aboveground part of the plant.

Careful selection of plant species is the determining factor for the success of phytostabilization. Plant characteristics and soil properties are both important for proper selection of the most suitable phytoagent. Native species which can survive in targeted contaminated soil are made a preferred choice in many phytostabilization processes (Solís-Dominguez et al., 2012). In addition, the plants appropriate for phytostabilization are grasses and fast-growing plants which can provide sufficient coverage with developed root systems to stabilize trace elements. Selected plants should be simple for further maintenance after establishment.

2.4 Bioenergy Crops and Phytostabilization Options

Bioenergy crops are promising candidates for phytostabilization of soils contaminated with trace elements due to their good ability to grow in contaminated and marginal soils (Pidlisnyuk et al., 2014). Below the main energy

crops proposed for application in phytoremediation process are characterized. It should be noted here that for the same energy crop, different results
have been observed. Differential response is due to multiple reasons, such as
soil type, soil amount (for example, small pot studies with insufficient amount
of soil), nature of soil contamination (field contaminated versus trace element
spiked soils, aged versus not aged, or fresh), and cleaning methods employed
for plant materials (i.e., inability to remove surface contaminations).

Arundo donax L. (Giant reed, Figure 2.2) is a perennial, tall, and upright
grass with highly efficient C_3 photosynthesis. It belongs to the *Poaceae* family.
The origin of Giant reed is not yet accurately defined; however, it is believed
to be native from Asia or the Mediterranean basin. The plant is tolerant to

FIGURE 2.2
Arundo donax, Manhattan, KS Nov 2020. Photo by Lawrence Davis.

unsuitable growing conditions, yet showing best growth in areas with good access to water (Angelini et al., 2005). Giant reed is cultivated well in moderate, subtropical, and tropical areas of both hemispheres (Herrera & Dudley, 2003). It grows between 6 and 8 m high, while in ideal conditions the height can exceed 10 m. It is a promising crop for energy production in the Mediterranean climate in Europe and Africa. The advantage is that the crop is adapted to long drying periods (Jeguirim & Trouvé, 2009; Zema et al., 2012). Plantings can last 12–15 years and may annually produce up to 60 t of dry biomass ha^{-1} in Central and Northern Italy (Angelini et al., 2005; Pilu et al., 2013). A considerable amount of research is reported in Table 2.2 about phytoremediation potential of *A. donax* in a variety of contaminated soils; however, results are mixed.

Panicum virgatum L. (Switchgrass) (Figure 2.3) is an upright, coarse, perennial C$_4$ grass. It belongs to the *Poaceae* family and originates in North America and Canada. The plant can be produced over an extensive geographic range, and the annual harvested biomass is up to 25 t of dry matter ha^{-1} (Parrish et al., 2012) while stem can grow up to 2.7 m. Switchgrass develops well in marginal soils and shows good results in both fine and coarse textured soils (Rinehart, 2006). The recommended seeding rate is 200–400 germinating seeds m^{-2}. Usually, the amount of seeds for one ha varies from 4 to 10 kg ha^{-1} (Moser & Vogel, 1995; Teel, 2003; Vassey et al., 1985; Vogel, 1987, 2000; Wolf & Fiske, 2009). Currently cultivation of switchgrass uses mainly hybrids. Some studies with trace elements are reported in Table 2.3.

TABLE 2.2

Phytoremediation Potential of *Arundo donax* L. to Different Trace Elements

Trace Element	Phytoremediation Potential	Reference
Cd	Rhizofiltration in hydroponics Phytoextraction Uptake potential from the media	Dürešová et al. (2014); Sagehashi et al. (2011) Barbafieri et al. (2011); Chierchia, (2011); Sabeen et al. (2013); Yang et al. (2012) Papazoglou et al. (2005); Papazoglou et al. (2007)
Cr	Accumulator	Fiorentino et al. (2013); Kausar et al. (2012)
Cu	Phytoextraction Rhizofiltration	Chierchia (2011); Elhawat et al. (2014) Bonanno (2012)
Hg	Uptakes and accumulates Hg in roots	Bonanno (2012)
Ni	Uptake potential from the media	Bonanno (2012); Papazoglou et al. (2005); Papazoglou et al. (2007)
Pb	Not efficient in removing Pb from the media Certain genotypes have phytoextraction potential; others are excluders	Barbafieri et al. (2011); Bonanno (2012) Sidella (2014)
Zn	Phytoextraction in hydroponics	Dürešová et al. (2014)

FIGURE 2.3
Switchgrass, (*Panicum virgatum*) courtesy of Professor John Dolman, K-State.

TABLE 2.3

Phytoremediation Potential of *Panicum virgatum* L. to Different Trace Elements

Trace Element	Phytoremediation Potential	References
Cd	Rhizofiltration Phytoextraction Accumulator	Abe et al. (2008); Chen et al. (2008); Sankaran & Ebbs (2007) Gerst (2014); Juang & Lee (2010) Chen et al. (2012)
Cr	Phytoextraction in hydroponics Potential remediator Rhizoextraction	Chen et al. (2012) Shahandeh & Hossner (2000) Li et al. (2011)
Hg	Not efficient	Gerst (2014)
Pb	Excluder Favorable for phytoremediation Not efficient as accumulator	Gleeson (2007) Johnson (2014) Żurek et al. (2013)
Zn	Accumulator Favorable for phytostabilization	Chen et al. (2012)

Pennisetum purpureum Schum. (Napier grass or Elephant grass) is a dense rhizomatous perennial C$_4$ grass which is often crossed with *P. americanum* to obtain a hybrid with better properties (Figure 2.4). It belongs to the *Poaceae* family and originates from sub-Saharan Africa; currently, it is widespread in tropical and subtropical regions. This species prefers areas with high precipitation; however, it also tolerates dry conditions due to a well-developed vigorous root system. The best growth is reported for deep, fertile loams, although it grows well on more marginal lands. *P. purpureum* is an aggressive grass able to grow rapidly, colonize new areas, and form dense thickets; moreover, the species is recognized globally as a very invasive grass. *Pennisetum purpureum*

FIGURE 2.4
Pennisetum purpureum Schum (Elephant grass).

grows forming thick clumps up to 1 m in diameter, with stems branched above, reaching 4–7 m height. Leaf length and width are around 100–120 cm and 1–5 cm, respectively (CABI, 2014). This plant can be used as a source for cellulosic bioenergy, fodder, cover material, bedding, and paper (Adekalu et al., 2007; Kabi et al., 2005). Positive results were recorded when it was used for treatment of waste sludge (Dhulap & Patil, 2014) and soil contaminated by hydrocarbons (Ayotamuno et al., 2006). Phytoremediation potentials of elephant grass for different trace elements are reported in Table 2.4.

Sida hermaphrodita L. Rusby (Virginia mallow) is a C_4, honey plant species (Figure 2.5) which belongs to *Malvaceae* family (mallows); it originated in North America. During the 1930s the plant was introduced to former USSR, and currently it can be found in all parts of Europe. Virginia Mallow is tolerant to extreme types of continental climate and can survive in cold conditions

TABLE 2.4

Phytoremediation Potential of *Pennisetum purpureum* Schum. to Different Trace
Elements

Trace Element	Phytoremediation Potential	References
Cd	Rhizofiltration	Lotfy et al. (2012)
	Phytoextraction	Abdel-Salam (2012); Zhang et al. (2010)
	Accumulator	Ogunkunle et al. (2014)
Cr	Excluder	Lotfy & Mostafa (2014)
	Accumulator in humid conditions	Ogunkunle et al. (2014)
Cu	Excluder	Ogunkunle et al. (2014); Yang et al. (2010)
Pb	Excluder	Xia (2004); Yang et al. (2010)
	Accumulator	Ogunkunle et al. (2014)
Zn	Excluder	Ogunkunle et al. (2014); Zhang et al. (2010)

FIGURE 2.5
Sida hermaphrodita a perennial dicot used for forage, or biomass production. With two harvests
per year it can serve for biogas production with a total annual harvest exceeding 26 Mg/ha.
This photo, courtesy of Professor Marek Bury, ZUT, Szczecin, Poland, shows regrowth after a
first harvest, alongside the crop maturing with flowers.

(even without snow at temperatures below –20°C) and dry conditions if the
average annual precipitation ranges between 400 and 500 mm. Height in full
maturity varies from 1 to 4 m, commonly reaches about 3 m (Borkowska &
Molas, 2012). Its life span is about 25 years (Kasprzyk et al., 2013), the annual
yield ranges from 15 to 20 t of dry matter ha⁻¹ when cultivated in clay loam soils
(Borkowska, 2007). In case of unfavorable conditions, the cultivation is often provided with the addition of sewage sludge; in this condition yield ranges from

none to 11 tons of dry matter ha^{-1} (Borkowska & Wardzinska, 2003). The well-developed root system allows it to efficiently use limited nutrients and water from marginal soils (Borkowska & Wardzinska, 2003). *S. hermaphrodita* grows well in stony or sandy soil with high yields, and best growth is reported for moderately humid areas. Hybrids and cultivars are mainly cultivated now because they have higher yields than the original species. The data about phytoremediation potential of Virginia mallow are summarized in Table 2.5.

Sorghum × drummondii Steud. (Sudan grass) is an annual, warm-season, fast growing plant (Figure 2.6), which belongs to *Poaceae* family and is a hybrid of *S. bicolor* and *S. arundinaceum*, and possesses C$_4$ photosynthesis. Sudan grass is originally from Southern Egypt and Sudan; in 1909 it was imported to the US where it began to be grown as a fodder species. The crop is widespread in South America, Australia, South Africa, Central and Northern Europe. It shows best growth in areas with average annual precipitation between 600 and 900 mm, nevertheless tolerates drought periods, and can be produced in all soil types (FAO, 2012). Recently *S. drummondii* has attracted interests due to its ability to remove trace elements from different media (Table 2.6). According to Pivetz (2001) it can absorb and accumulate Co. Application of microbial inoculants improved the remediation process (Shim et al., 2014). Utilization of certain mycorrhizal fungi which form a symbiosis with *S. × drummondii* can increase the accumulation of trace elements from the contaminated soil (Gaur & Adholeya, 2004).

The Miscanthus genus belongs to the *Poaceae* (or grasses) family (Figure 2.7). It has C$_4$ photosynthesis with high water and nutrient-use efficiency and cold tolerance (Chung & Kim, 2012). The genus can be found within high lawns of

TABLE 2.5

Phytoremediation Potential of *Sida hermaphrodita* L. to Different Trace Elements

Trace Element	Phytoremediation Potential	References
Cd	Phytoextraction	Ociepa (2011)
	Rhizofiltration	Antonkiewicz & Jasiewicz (2002)
	Potential accumulator	Borkowska et al. (2001); Krzywy-Gawrońska (2012)
Cr	Efficiently removes Cr from the media	Borkowska et al. (2001)
Cu	Efficiently removes Cu from the media	Borkowska et al. (2001)
		Antonkiewicz & Jasiewicz (2002);
	Rhizoextraction	Krzywy-Gawrońska (2012)
Ni	Efficient in phytoextraction	Krzywy-Gawrońska (2012)
		Antonkiewicz & Jasiewicz (2002)
	Rhizoextraction	
Pb	Phytoextraction	Krzywy-Gawrońska (2012)
	Accumulator	Kocoń & Matyka (2012)
Zn	Phytoextraction	Krzywy-Gawrońska (2012)
	Accumulator	Borkowska et al. (2001)

FIGURE 2.6
Sudan-Sorghum forage hybrid at the USDA Plant Materials Center, Ashland Bottoms, Manhattan, KS, August 2013. Photo courtesy of Mark Janzen, Natural Resources Conservation Services, USDA.

TABLE 2.6

Phytoremediation Potential of *Sorghum × drummondii* Steud. to Different Trace Elements

Trace Element	Phytoremediation Potential	References
Cd	Excluder Rhizoextraction	Angelova et al. (2011); López-Chuken and Young (2005); Zwonitzer et al. (2003) Da-lin et al. (2011)
Cu	Excluder	Angelova et al. (2011); Tari et al. (2013)
Pb	Suitable for phytoremediation Chemically induced remediation Excluder	Murányi and Ködöböcz (2008) Zhuang et al. (2009) Angelova et al. (2011)
Zn	Excluder	Angelova et al. (2011)

Eastern Asia, from tropics and subtropics to Pacific islands, warm temperate regions, and subarctic areas. Wide adaptability to various environmental factors makes *M. × giganteus* a sterile triploid hybrid of *M. sinensis* and *M. sacchariflorus*, suitable for cultivating in different European and Northern American climatic conditions (Greef & Deuter, 1993, See Chapter 5 for further information about breeding of various hybrid Miscanthus and different species properties). Plant biomass annual yield ranges from 8 to 45 t dry biomass ha^{-1} (Bilandžija, 2015; Heaton et al., 2008; Lewandowski et al., 2000; Maughan et al., 2012; Miguez et al., 2008) with long sustainable productivity

FIGURE 2.7
Miscanthus at Fort Riley, site, KS. USA. (Photo: Zafer Alasmary, Kansas State University, KS, USA.)

after the establishment of up to 20 years, or more. *M. × giganteus* has a high carbon sequestration capacity as well because of its dense rhizome and root system (Chung & Kim, 2012). Commonly, the initial planting density per ha is between 10,000 and 13,000 plants. The main characteristics of Miscanthus as a prospective biofuel crop are as follows: exceptional cultivation adaptation in different climatic and pedological conditions, possibility to grow in soils of inferior quality, high dry matter yields, high energy value, exceptional resistance to diseases and pests, and low demand for nutrients. As a natural sterile hybrid, invasive spread is much lower in comparison with some other energy crops (Bilandžija, 2014).

2.5 *M. × giganteus* as an Effective Phytoagent

Plants have to meet certain requirements for application in phytoremediation of contaminated sites: to be resistant to pests, plant diseases, and contaminants of different origin; to be able to grow at the same site long term. *M.* ×*giganteus* has all of these properties (Clifton-Brown et al., 2008; Dauber et al., 2010). The plant is among the more promising bioenergy crops due to high yield, water and nutrient use efficiency, and lignocellulose content. Lewandowski et al. (2016) pointed that *M.* ×*giganteus* has only weak ability to absorb contaminants; therefore, it can be cultivated in contaminated soils while obtaining relatively clean biomass.

The selection of Miscanthus for phytomanagement of different contaminated areas with simultaneous production of alternative energy source is getting popular (Amougou et al., 2011; Dubis et al., 2019; Kołodziej et al., 2016; Kvak et al., 2018; Lewandowski et al., 2005; Matyka & Kuś, 2016; Roik et al., 2019; Tryboi, 2018). Thus, because Miscanthus covers 15% of the total bioenergy plant market in some Eastern European countries (Geletukha et al., 2016), can stabilize and accumulate some trace elements, absorb and degrade contaminants of organic origin, facilitate carbon deposition, and improve physico-chemical properties of soil, this plant may be successfully applied for restoration of postmining (Kharytonov et al., 2019; Nurzhanova et al., 2019; Pidlisnyuk et al., 2020a, b) and postmilitary (Alasmary, 2020; Pidlisnyuk et al., 2018, 2019) sites (Table 2.7). Miscanthus brings ecological benefits with economic profit due to using biomass for energy and bioproducts.

2.5.1 Miscanthus Tolerance to Metals and Removal Capacity

Miscanthus' ability to grow in soils contaminated by trace elements (Alasmary, 2020; Pidlisnyuk et al., 2019; Wilkins, 1997) is determined by the following factors (Wang et al., 2020):

TABLE 2.7

Phytoremediation Potential of *M.* ×*giganteus* to Different Trace Elements

Trace Element	Phytoremediation Potential	Reference
Cd	Rhizoextraction	Arduini et al. (2006)
	Excluder	Fernando and Oliveira (2004)
	Phytostabilization	Nsanganwimana et al. (2015)
Cr	Rhizoextraction	Arduini et al. (2006)
	Excluder	Fernando and Oliveira (2004)
Cu	Excluder	Fernando and Oliveira (2004)
Hg	Excluder	Fernando and Oliveira (2004)
Ni	Excluder	Fernando and Oliveira (2004)
Pb	Excluder	Pavel et al. (2014)
	Excluder	Fernando and Oliveira (2004)
	Phytostabilization	Nsanganwimana et al. (2015)
Zn	Accumulator	Pogrzeba et al. (2013)
	Excluder	Fernando and Oliveira (2004)
	Phytostabilization	Nsanganwimana et al. (2015)

a. Miscanthus root systems are large and well-developed, and plant metabolism is vigorous. Also, carbon-containing compounds released by plant roots supply the microorganisms located in rhizosphere (Rhizobacteria) with nutrients and organic acids (Hromádko et al., 2014; Zgorelec et al., 2020). Such acids can suppress trace element toxicity. Guo et al. (2017) reported that under Cd stress *M. sacchariflorus* roots secreted malate which mitigated Cd toxicity for the plant, by reducing its absorption.

b. Antioxidant and photosynthetic activities of Miscanthus are well developed. The antioxidant defense system plays a crucial role in plant stress response. Along with the ability to mitigate stress-induced disturbances, it can serve as an indicator of trace element toxicity-induced stress. An increase of malondialdehyde content reflects Cr stress (Jiang et al., 2018). Significant increases in chlorophyll content, superoxide dismutase, and peroxidase activities are observed in *M. floridulus* and *M. sacchariflorus* growing in soil slightly contaminated by Pb, Zn, or Cd (Zhang et al., 2015). *M. × giganteus* behavior was similar under Pb and Zn stresses (Nurzhanova et al., 2019).

c. The Miscanthus rhizosphere contains many microbial colonies that participate in plant–soil interaction (Wang et al., 2020). Schmidt et al. (2018) reported that plant inoculation with bacteria and fungi isolated from different Miscanthus species' rhizospheres improved plant growth. Firmin et al. (2015) obtained the same result after inoculating *M. × giganteus* by *Funneliformis mosseae*. Inoculation of Miscanthus rhizomes with plant growth promoting bacteria increased the biomass by ~77% after first vegetation season in postmining soil contaminated by trace elements (Pidlisnyuk et al., 2020a; Pranaw et al., 2020).

2.5.2 Changes in Soil Parameters Induced by Miscanthus Phytoremediation

Miscanthus planting in contaminated soils can increase soil carbon content, enhance aggregate stability, and improve water-holding capacity. The improving effect of the plant on soil physiochemical properties is mainly attributed to decomposition of underground organs and litter of root residuals in soil (Wang et al., 2020). McCalmont et al. (2015) showed that the decomposition of litter and underground organs of Miscanthus provides a large amount of organic carbon to soil, which increases soil organic matter, promotes soil nutrient cycling, improves the texture, structure, and water-holding soil capacity, and reduces soil nutrient loss.

Miscanthus was cultivated in marginal soil with the application of soil amendments, biochar, and biosolid, that enhanced the abundance of humus and mycorrhizal fungi, and improved soil fertility and hydraulic properties. Biosolids exerted the most pronounced effect (Allami et al., 2019).

Zhang et al. (2020) applied mixed planting of Miscanthus, Masson pine, and Bamboo for phytoremediation of a mining area. After 3 years of cultivation, the abundance of the mixed vegetation was tremendously higher, and microflora in remediated soil was larger in comparison with the control one. Miscanthus significantly reduces the release of NO_x and increases the absorbing capacity for CH_4. The net release of greenhouse gases was reduced to an extent of 4.08 t CO_2-eq per hectare per year (Mi et al., 2018).

In the long run during Miscanthus cultivation, soil carbon sequestration could be significantly different (Holder et al., 2019). During one life cycle (15 years), Miscanthus may release twice the amount of greenhouse gas in comparison to permanent grassland. Compared to the grassland soils, the surface soils of Miscanthus fields tend to have a risk of acidification due to higher concentrations of P and K (Hu et al., 2018). Therefore, when evaluating the impacts of Miscanthus cultivation, soil characteristics and soil organic carbon stability should be taken into consideration in the long-term perspective.

2.6 Miscanthus Phytotechnology in Action

2.6.1 *M.* × *giganteus* Application for Phytoremediation of Trace Elements' Contaminated Mining Soil, Tekeli, Kazakhstan

The research soil was sampled around the Tekeli Mining and Processing Complex of "Kazzinc", Kazakhstan. Soil was contaminated by trace elements in concentrations that exceeded the maximum permissible levels in Kazakhstan, i.e., the exceeding for Pb was in 29 times, As – 5 times, Zn – 11 times, Sr – 22 times, Cu – 13 times. Research soil belonged to saline and sandy types; its pH (water) ranged from 8.3 to 9.9 (Nurzhanova et al., 2019).

The experiment lasted two vegetation seasons. During the experiment soil and *M.* × *giganteus* tissues (root, stems, and leaves) were analyzed for the content of eleven trace elements: As, Pb, Zn, Co, Ni, Cr, Cu, Sr, Mn, V, and U. These elements mainly accumulated in the *M.* × *giganteus* root system. Accumulation of Zn, Sr, and Mn was higher than others (Figure 2.8). *M.* × *giganteus* behaved as an excluder (BCF and TLF values are lower than 1) preferentially accumulating the observed trace elements in roots; however, in relation to Mn, Sr, and Zn the plant acted as extractor (BCF < 1 and TLF ≥ 1) (Nurzhanova et al., 2019).

2.6.2 *M.* × *giganteus* Application for Phytoremediation of Post-Industrial Soil Contaminated with Trace Elements, Bakar, Croatia

This experiment was based on monitoring of *M.* × *giganteus* phytoremediation potential in relation to post-industrial soil of Rijeka-Bakar industrial zone, Croatia, which was contaminated by different trace elements (Pidlisnyuk et al., 2020b). The biomass parameters and concentrations of Ti, Mn, Fe, Cu,

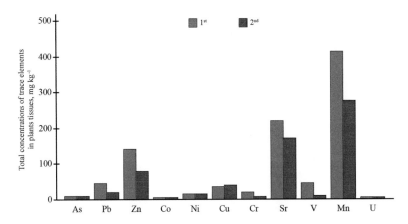

FIGURE 2.8
Element accumulations in *M. ×giganteus* tissues during two vegetation seasons

Zn, As, Sr, and Mo in stems, leaves, and roots were analyzed in each of three vegetation seasons at harvest. The following categories were used as features affecting the phytoremediation process: the difference in trace elements distribution along the plant; trace element concentrations in the researched soil; different regimes of trace elements absorption by roots and transmigration to plant organs; and vegetation season. Results of the statistical analysis (Pidlisnyuk et al., 2020b) showed that the main factor was trace element organ distribution which was essential for Ti, Fe, and Cu. The difference in trace element concentration in soil significantly correlated with Zn and Mo was essentially lower with As, Sr, and Mn, while for Ti and Cu correlation was not detected. The impact of the combined effect of two factors (trace elements organ distribution and difference in trace element concentration in soil) was detected for two elements: more prominent for Cu and smaller for Ti.

The plants organs (variable "Zone") mainly affect the trace elements concentration variations, i.e., accumulation in specific plants part: it was the most essential for Ti, Fe, and Cu and the smallest for Mn. The second factor (trace elements concentration in soil – variable "Experiment") was the most essential for Zn and Mo; however, much less for As, Sr, and Mn; limited for Fe; and was not observed for Ti and Cu. The combined effect of the above two factors was detected for two elements: higher for Cu and lower for Ti (Figure 2.9).

2.6.3 Field Study Results, Fort Riley, Kansas, USA

In a field study established in 2016, on an US Army reservation in Fort Riley, KS, Miscanthus was planted in an area with soil total Pb concentration ranging from 900 to 1500 mg kg^{-1} and near-neutral soil pH. Five treatments were evaluated: (i) control without tillage with existing vegetation; (ii) no-tillage, no additional amendments planted with Miscanthus; (iii) tilled soil, no additional amendments planted with Miscanthus; (iv) tilled soil amended with triple

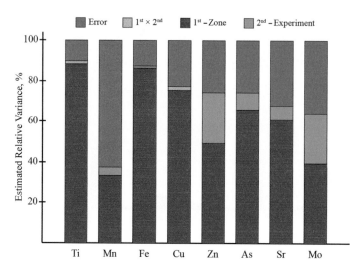

FIGURE 2.9
The components of the element concentration variation (after Box-Cox transformation) depended on plant organs (variable 1 – "Zone"), experiment treatment (variable 2 – "Experiment"), and its interaction (1*2) (with vegetation duration as a covariate). Notes: Zone – the effect of the plant organs (roots, leaves, stems), experiment – the effect of the experiment treatments (level 1–5), 1*2 – the interaction effects of the Zone and Experiment.

superphosphate (at 5:3 Pb:P molar ratio) planted with Miscanthus; and (v) tilled soil amended with organic P source (class B biosolids applied at 45 Mg ha⁻¹ air-dry weight basis) planted with Miscanthus. Results from 2016 to 2018 showed that one-time addition of soil amendments to Pb-contaminated soil supports establishing and stabilizing Miscanthus, increasing biomass yield as well as reducing phytoavailability and bioaccessibility of Pb (as measured by physiological-based extraction test procedure developed by Ruby et al. (1996) and modified by Medlin (1997)). Moreover, biosolids-treated plots showed improved soil enzyme activities, organic carbon, and microbial biomass (Alasmary et al., 2020). X-ray absorption spectroscopy results indicated pyromorphite, Pb associated with Fe minerals, and Pb adsorbed to humic acid were the dominant Pb species in P-amended and nonamended soils (Unpublished data, Alasmary and Hettiarachchi). The results suggest that Miscanthus can be grown successfully in Pb-contaminated military site soils combined with soil amendments, while minimizing the associated environmental risks.

2.7 Conclusions

Using energy crops in phytostabilization of soils contaminated with trace elements is one of the green technologies that provides ecological, economic, and social solutions for contaminated areas, while meeting energy needs and mitigating climate change. The classification of plant species into corresponding groups (hyperaccumulator, accumulator, indicator, excluder) is complex as different conditions (soil type, pH, climate, location, media, plant properties, choice of cultivars, etc.) can influence plant uptake of trace elements and there is no universal approach. Miscanthus proved its ability to grow and to remediate soils contaminated by trace elements with sufficient biomass yield. Besides the obvious advantages of Miscanthus as a phytoremediation agent, its cultivation on the marginal and slightly contaminated lands can improve the soil biological parameters, such as basal respiration, microbial biomass carbon, fluorescein diacetate hydrolytic activity, other enzymatic activities, and simultaneously prevent soil and water erosion. The application of Miscanthus in phytomanagement can improve soil health and help to supply biomass for utilization to obtain energy or bioproducts. Utilization of Miscanthus for phytomanagement of differently contaminated soils is a prospective green technology with potential widespread commercial feedback.

References

Abdel-Salam, M. (2012). Chemical and phyto-remediation of clayey and sandy textured soils polluted with cadmium. *American-Eurasian Journal of Agricultural & Environmental Sciences*, 12(6), 689–693. https://doi.org/10.5829/idosi. aejaes.2012.12.06.1801.

Abe, T., Fukami, M., & Ogasawara, M. (2008). Cadmium accumulation in the shoots and roots of 93 weed species. *Soil Science & Plant Nutrition*, 54(4), 566–573. https://doi.org/10.1111/j.1747-0765.2008.00288.x.

Adekalu, K. O., Olorunfemi, I. A., & Osunbitan, J. A. (2007). Grass mulching effect on infiltration, surface runoff and soil loss of three agricultural soils in Nigeria. *Bioresource Technology*, 98(4), 912–917. https://doi.org/10.1016/j. biortech.2006.02.044.

Adriano, D. (2001). Bioavailability of trace metals. In Domy C. Adriano (Ed.), Trace Elements in the Terrestrial Environment (2nd Edition, pp. 61–89), Springer-Verlag, New York.

Alasmary, Z. (2020). *Laboratory- to field-scale investigations to evaluate phosphate amendments and Miscanthus for phytostabilization of lead-contaminated military sites* [PhD, Kansas State University]. https://krex.k-state.edu/dspace/ handle/2097/40676.

Alasmary, Z., Todd, T., Hettiarachchi, G. M., Stefanovska, T., Pidlisnyuk, V., Roozeboom, K., Erickson, L., Davis, L., & Zhukov, O. (2020). Effect of soil treatments and amendments on the nematode community under Miscanthus growing in a lead contaminated military site. *Agronomy*, 10(11), 1727. https:// doi.org/10.3390/agronomy10111727.

Allami, M., Oustriere, N., Gonzales, E., & Burken, J. G. (2019). Amendment-assisted revegetation of mine tailings: Improvement of tailings quality and biomass production. *International Journal of Phytoremediation*, 21(5), 425–434. https://doi.org/ 10.1080/15226514.2018.1537249.

Amougou, N., Bertrand, I., Machet, J.-M., & Recous, S. (2011). Quality and decomposition in soil of rhizome, root and senescent leaf from *Miscanthus × giganteus*, as affected by harvest date and N fertilization. *Plant and Soil*, 338(1), 83–97. https://doi.org/10.1007/s11104-010-0443-x.

Angelini, L. G., Ceccarini, L., & Bonari, E. (2005). Biomass yield and energy balance of giant reed (*Arundo donax L.*) cropped in central Italy as related to different management practices. *European Journal of Agronomy*, 22(4), 375–389. https://doi.org/10.1016/j.eja.2004.05.004.

Angelova, V., Ivanova, R., Delibaltova, V., & Ivanov, K. (2011). Use of sorghum crops for in situ phytoremediation of polluted soils. *Journal of Agricultural Science and Technology. A*, 1(5), 693–702.

Antonkiewicz, J., & Jasiewicz, C. (2002). The use of plants accumulating heavy metals for detoxification of chemically polluted soils. *Electronic Journal of Polish Agricultural Universities*, 5(1), 121–143.

Arduini, I., Ercoli, L., Mariotti, M., & Masoni, A. (2006). Response of miscanthus to toxic cadmium applications during the period of maximum growth. *Environmental and Experimental Botany*, 55(1), 29–40. https://doi.org/10.1016/j.envexpbot.2004.09.009.

Attanayake, C. P., Hettiarachchi, G. M., Harms, A., Presley, D., Martin, S., & Pierzynski, G. M. (2014). Field evaluations on soil plant transfer of lead from an urban garden soil. *Journal of Environmental Quality*, 43(2), 475–487. https://doi.org/10.2134/jeq2013.07.0273.

Ayotamuno, J. M., Kogbara, R. B., & Egwuenum, P. N. (2006). Comparison of corn and elephant grass in the phytoremediation of a petroleum-hydrocarbon-contaminated agricultural soil in Port Harcourt, Nigeria. *Journal of Food Agriculture and Environment*, 4(3/4), 218.

Baker, A. J. M. (1981). Accumulators and excluders strategies in the response of plants to heavy metals. *Journal of Plant Nutrition*, 3(1–4), 643–654. https://doi.org/10.1080/01904168109362867.

Baker, A. J., & Brooks, R. (1989). Terrestrial higher plants which hyperaccumulate metallic elements. A review of their distribution, ecology and phytochemistry. *Biorecovery*, 1(2), 81–126.

Barbafieri, M., Dadea, C., Tassi, E., Bretzel, F., & Fanfani, L. (2011). Uptake of heavy metals by native species growing in a mining area in Sardinia, Italy: Discovering native flora for phytoremediation. *International Journal of Phytoremediation*, 13(10), 985–997. https://doi.org/10.1080/15226514.2010.549858.

Berti, W. R., & Cunningham, S. D. (1997). In-place inactivation of Pb in Pb-contaminated soils. *Environmental Science & Technology*, 31(5), 1359–1364. https://doi.org/10.1021/es960577+.

Berti, W. R., & Cunningham, S. D. (2000). Phytostabilization of metals. In Ilya Raskin & Burt D. Ensley (Eds.), *Phytoremediation of Toxic Metals: Using Plants to Clean up the Environment* (pp. 71–88), Wiley, New York.

Bilandžija, N. (2014). Perspective and potential use of Miscanthus × giganteus culture in Croatia [Perspektiva i potencijal korištenja kulture Miscanthus × giganteus u Republici Hrvatskoj]. *Environmental Engineering - Inženjerstvo okoliša*, 1(2), 81–87. https://hrcak.srce.hr/index.php?show=clanak&id_clanak_jezik=195970.

Bilandžija, N. (2015). *The potential of Miscanthus × giganteus species as an energy crop in different technological and environmental conditions [Potencijal vrste Miscanthus × giganteus kao energetske kulture u različitim tehnološkim i agroekološkim uvjetima]* [PhD, University of Zagreb].

Bonanno, G. (2012). *Arundo donax* as a potential biomonitor of trace element contamination in water and sediment. *Ecotoxicology and Environmental Safety*, 80, 20–27. https://doi.org/10.1016/j.ecoenv.2012.02.005.

Borkowska, H. (2007). Yields of Virginia fanpetals and willow on good wheat soil complex. *Fragmenta Agronomica*, 2, 41–47.

Borkowska, H., Jackowska, I., Piotrowski, J., Styk, B., Piotrowski, J., & Styk, B. (2001). Suitability of cultivation of some perennial plant species on sewage sludge. *Polish Journal of Environmental Studies*, 10(5), 379–382.

Borkowska, H., & Molas, R. (2012). Two extremely different crops, Salix and Sida, as sources of renewable bioenergy. *Biomass and Bioenergy*, 36, 234–240. https://doi.org/10.1016/j.biombioe.2011.10.025.

Borkowska, H., & Wardzinska, K. (2003). Some effects of Sida hermaphrodita R. cultivation on sewage sludge. *Polish Journal of Environmental Studies*, 12(1), 119–122.

Brown, S., Chaney, R. L., Hallfrisch, J. G., & Xue, Q. (2003). Effect of biosolids processing on lead bioavailability in an urban soil. *Journal of Environmental Quality*, 32(1), 100–108. https://doi.org/10.2134/jeq2003.1000.

CABI. (2014). *Invasive Species Compendium*. Centre for Agriculture and Bioscience International.

Chen, B.-C., Lai, H.-Y., & Juang, K.-W. (2008). Model evaluation of plant metal content and biomass yield for the phytoextraction of heavy metals by switchgrass. *Ecotoxicology and Environmental Safety*, 80, 393–400. https://doi.org/10.1016/j.ecoenv.2012.04.011.

Chen, B.-C., Lai, H.-Y., & Juang, K.-W. (2012). Model evaluation of plant metal content and biomass yield for the phytoextraction of heavy metals by switchgrass. *Ecotoxicology and Environmental Safety*, 80, 393–400. https://doi.org/10.1016/j.ecoenv.2012.04.011.

Chierchia, A. (2011). *L'Arundo donax per la bonifica dei suoli contaminati da metalli pesanti* [Master, University of Napoli "Federico II"].

Chung, J.-H., & Kim, D.-S. (2012). Miscanthus as a potential bioenergy crop in East Asia. *Journal of Crop Science and Biotechnology*, 15(2), 65–77. https://doi.org/10.1007/s12892-012-0023-0.

Clifton-Brown, J., Chiang, Y.-C., & Hodkinson, T. R. (2008). Miscanthus: Genetic resources and breeding potential to enhance bioenergy production. In W. Vermerris (Ed.), *Genetic Improvement of Bioenergy Crops* (pp. 295–308), Springer, New York. https://doi.org/10.1007/978-0-387-70805-8_10.

Conesa, H. M., Evangelou, M. W. H., Robinson, B. H., & Schulin, R. (2012, January 4). A critical view of current state of phytotechnologies to remediate soils: Still a promising tool? [Review Article]. *The Scientific World Journal*. https://doi.org/10.1100/2012/173829.

Da-lin, L., Kai-qi, H., Jing-jing, M., Wei-wei, Q., Xiu-ping, W., & Shu-pan, Z. (2011). Effects of cadmium on the growth and physiological characteristics of sorghum plants. *African Journal of Biotechnology*, 10(70), 15770–15776. https://doi.org/10.4314/ajb.v10i70.

Dauber, J., Jones, M. B., & Stout, J. C. (2010). The impact of biomass crop cultivation on temperate biodiversity. *GCB Bioenergy*, 2(6), 289–309. https://doi.org/10.1111/j.1757-1707.2010.01058.x.

Defoe, P. P., Hettiarachchi, G. M., Benedict, C., & Martin, S. (2014). Safety of gardening on lead- and arsenic-contaminated urban brownfields. *Journal of Environmental Quality*, 43(6), 2064–2078. https://doi.org/10.2134/jeq2014.03.0099.

Dhulap, V., & Patil, S. (2014). Removal of pollutants from sewage through constructed wetland using *Pennisetum purpureium*. *European Educational Research Journal*, 2(1), 543–558.

Dubis, B., Jankowski, K. J., Załuski, D., Bórawski, P., & Szempliński, W. (2019). Biomass production and energy balance of Miscanthus over a period of 11 years: A case study in a large-scale farm in Poland. *GCB Bioenergy*, 11(10), 1187–1201. https://doi.org/10.1111/gcbb.12625.

Dürešová, Z., Šuňovská, A., Horník, M., Pipíška, M., Gubišová, M., Gubiš, J., & Hostin, S. (2014). Rhizofiltration potential of *Arundo donax* for cadmium and zinc removal from contaminated wastewater. *Chemical Papers*, 68(11), 1452–1462. https://doi.org/10.2478/s11696-014-0610-2.

Elhawat, N., Alshaal, T., Domokos-Szabolcsy, É., El-Ramady, H., Márton, L., Czakó, M., Kátai, J., Balogh, P., Sztrik, A., Molnár, M., Popp, J., & Fári, M. G. (2014). Phytoaccumulation potentials of two biotechnologically propagated ecotypes of *Arundo donax* in copper-contaminated synthetic wastewater. *Environmental Science and Pollution Research*, 21(12), 7773–7780. https://doi.org/10.1007/s11356-014-2736-8.

Evangelou, M. W. H., Papazoglou, E. G., Robinson, B. H., & Schulin, R. (2015). Phytomanagement: Phytoremediation and the production of biomass for economic revenue on contaminated land. In A. A. Ansari, S. S. Gill, R. Gill, G. R. Lanza, & L. Newman (Eds.), *Phytoremediation: Management of Environmental Contaminants* (Vol. 1, pp. 115–132). Springer International Publishing, New York. https://doi.org/10.1007/978-3-319-10395-2_9.

FAO. (2012). *The Grassland Index*. Food and Agriculture Organization. https://web.archive.org/web/20170102172754/http://www.fao.org/ag/AGp/agpc/doc/gbase/mainmenu.htm

Fernando, A., & Oliveira, J. (2004). *Effects on growth, productivity and biomass quality of Miscanthus × giganteus of soils contaminated with heavy metals* [2nd World Conference on Biomass for Energy, Industry and Climate Protection], 10–14 May, 2004, Rome, Italy. https://moodle.fct.unl.pt/pluginfile.php/92027/mod_resource/content/0/Fernando_e_Oliveira_2004.pdf. 10–14.

Fiorentino, N., Fagnano, M., Adamo, P., Impagliazzo, A., Mori, M., Pepe, O., Ventorino, V., & Zoina, A. (2013). Assisted phytoextraction of heavy metals: Compost and Trichoderma effects on giant reed (*Arundo donax* L.) uptake and soil N-cycle microflora. *Italian Journal of Agronomy*, 8(4), 244–254.

Firmin, S., Labidi, S., Fontaine, J., Laruelle, F., Tisserant, B., Nsanganwimana, F., Pourrut, B., Dalpé, Y., Grandmougin, A., Douay, F., Shirali, P., Verdin, A., & Lounès-Hadj Sahraoui, A. (2015). Arbuscular mycorrhizal fungal inoculation protects *Miscanthus × giganteus* against trace element toxicity in a highly metal-contaminated site. *Science of the Total Environment*, 527–528, 91–99. https://doi.org/10.1016/j.scitotenv.2015.04.116.

Gaur, A., & Adholeya, A. (2004). Prospects of arbuscular mycorrhizal fungi in phytoremediation of heavy metal contaminated soils. *Current Science*, 86(4), 528–534. https://www.jstor.org/stable/24107905.

Geletukha, G. G., Zheliezna, T. A., Tryboi, O. V., & Bashtovyi, A. I. (2016). Analysis of criteria for the sustainable development of bioenergy. *UABio Position Paper, 17*, 30. https://doi.org/10.31472/ihe.6.2016.07.

Gerst, E. A. (2014). *A novel approach to use second generation biofuel crop plants (Camelina sativa, Miscanthus × giganteus, and Panicum virgatum) to remediate abandoned mine lands in Pennsylvania* [Master, The Pennsylvania State University]. https://etda.libraries.psu.edu/catalog/21365.

Gleeson, A. M. (2007). *Phytoextraction of lead from contaminated soil by Panicum virgatum L. (switchgrass) and associated growth responses* [Master, Queen's University].

Greef, J., & Deuter, M. (1993). Syntaxonomy of *Miscanthus × giganteus* Greef et Deu. *Angewandte Botanik*, 67(3/4), 87–90.

Gudichuttu, V. (2014). *Phytostabilization of multi-metal contaminated mine waste materials: Long-term monitoring of influence of soil amendments on soil properties, plants, and biota and the avoidance response of earthworms* [Thesis, Kansas State University]. https://krex.k-state.edu/dspace/handle/2097/16989.

Guo, H., Wu, Y., Hong, C., Chen, H., Chen, X., Zheng, B., Jiang, D., & Qin, W. (2017). Enhancing digestibility of Miscanthus using lignocellulolytic enzyme produced by Bacillus. *Bioresource Technology*, 245, 1008–1015. https://doi.org/10.1016/j.biortech.2017.09.034.

Heaton, E. A., Dohleman, F. G., & Long, S. P. (2008). Meeting US biofuel goals with less land: The potential of Miscanthus. *Global Change Biology*, 14(9), 2000–2014. https://doi.org/10.1111/j.1365-2486.2008.01662.x.

Herrera, A. M., & Dudley, T. L. (2003). Reduction of riparian arthropod abundance and diversity as a consequence of giant reed (*Arundo donax*) invasion. *Biological Invasions*, 5(3), 167–177. https://doi.org/10.1023/A:1026190115521.

Hettiarachchi, G. M., Pierzynski, G. M., & Ransom, M. D. (2000). In situ stabilization of soil lead using phosphorus and manganese oxide. *Environmental Science & Technology*, 34(21), 4614–4619. https://doi.org/10.1021/es001228p.

Holder, A. J., Clifton-Brown, J., Rowe, R., Robson, P., Elias, D., Dondini, M., McNamara, N. P., Donnison, I. S., & McCalmont, J. P. (2019). Measured and modelled effect of land-use change from temperate grassland to Miscanthus on soil carbon stocks after 12 years. *GCB Bioenergy*, 11(10), 1173–1186. https://doi.org/10.1111/gcbb.12624.

Hromádko, L., Vranová, V., Techer, D., Laval-Gilly, P., Rejšek, K., Formánek, P., & Falla, J. (2014). Composition of root exudates of *Miscanthus × giganteus* Greef et Deu. *Acta Universitatis Agriculturae et Silviculturae Mendelianae Brunensis*, 58(1), 71–76.

Hu, Y., Schäfer, G., Duplay, J., & Kuhn, N. J. (2018). Bioenergy crop induced changes in soil properties: A case study on Miscanthus fields in the Upper Rhine Region. *PLoS One*, 13(7), e0200901. https://doi.org/10.1371/journal.pone.0200901.

Hunt, A. J., Anderson, C. W. N., Bruce, N., García, A. M., Graedel, T. E., Hodson, M., Meech, J. A., Nassar, N. T., Parker, H. L., Rylott, E. L., Sotiriou, K., Zhang, Q., & Clark, J. H. (2014). Phytoextraction as a tool for green chemistry. *Green Processing and Synthesis*, 3(1), 3–22. https://doi.org/10.1515/gps-2013-0103.

Jeguirim, M., & Trouvé, G. (2009). Pyrolysis characteristics and kinetics of *Arundo donax* using thermogravimetric analysis. *Bioresource Technology*, 100(17), 4026–4031. https://doi.org/10.1016/j.biortech.2009.03.033.

Jiang, H., Zhao, X., Fang, J., & Xiao, Y. (2018). Physiological responses and metal uptake of Miscanthus under cadmium/arsenic stress. *Environmental Science and Pollution Research*, 25(28), 28275–28284. https://doi.org/10.1007/s11356-018-2835-z.

Johnson, D. (2014). *Induced phytoextraction of lead from contaminated urban soil through manipulation of rhizosphere and plant biogeochemical functions in switch-grass (Panicum virgatum)* [Master of Science in Integrative Biology Theses, Kennesaw State University]. https://digitalcommons.kennesaw.edu/integrbiol_etd/2.

Juang, K.-W., & Lee, D. Y. (2010). *Prediction of cadmium removal from highly contaminated soils by phytoextraction with different switchgrass cultivars*, 19th World Congress of Soil Science, Soil Solutions for a Changing World, Brisbane, Australia.

Kabi, F., Bareeba, F. B., Havrevoll, Ø., & Mpofu, I. D. T. (2005). Evaluation of protein degradation characteristics and metabolisable protein of elephant grass (*Pennisetum purpureum*) and locally available protein supplements. *Livestock Production Science*, 95(1), 143–153. https://doi.org/10.1016/j.livprodsci.2004.12.013.

Kasprzyk, A., Leszczuk, A., Domaciuk, M., & Szczuka, E. (2013). Stem morphology of the *Sida hermaphrodita* (L.) Rusby (Malvaceae). *Modern Phytomorphology*, 4, 25–25.

Kausar, S., Mahmood, Q., Raja, I. A., Khan, A., Sultan, S., Gilani, M. A., & Shujaat, S. (2012). Potential of *Arundo donax* to treat chromium contamination. *Ecological Engineering*, 42, 256–259. https://doi.org/10.1016/j.ecoleng.2012.02.019.

Kharytonov, M., Pidlisnyuk, V., Stefanovska, T., Babenko, M., Martynova, N., & Rula, I. (2019). The estimation of *Miscanthus × giganteus*' adaptive potential for cultivation on the mining and post-mining lands in Ukraine. *Environmental Science and Pollution Research*, 26(3), 2974–2986.

Kocoń, A., & Matyka, M. (2012). Phytoextractive potential of *Miscanthus × giganteus* and *Sida hermaphrodita* growing under moderate pollution of soil with Zn and Pb. *Journal of Food, Agriculture and Environment*, 10(2 part 3), 1253–1256. https://www.cabdirect.org/cabdirect/abstract/20123224099.

Kołodziej, B., Antonkiewicz, J., & Sugier, D. (2016). *Miscanthus × giganteus*' as a biomass feedstock grown on municipal sewage sludge. *Industrial Crops and Products*, 81, 72–82. https://doi.org/10.1016/j.indcrop.2015.11.052.

Krzywy-Gawrońska, E. (2012). The effect of industrial wastes and municipal sewage sludge compost on the quality of virginia fanpetals (*Sida hermaphrodita* Rusby) biomass Part 1. Macroelements content and their uptake dynamics. *Polish Journal of Chemical Technology*, 14(2), 9–15. https://doi.org/10.2478/v10026-012-0064-7.

Kvak, V., Stefanovska, T., Pidlisnyuk, V., Alasmary, Z., & Kharytonov, M. (2018). The long-term assessment of *Miscanthus × giganteus* cultivation in the Forest-Steppe zone of Ukraine. *INMATEH-Agricultural Engineering*, 54(1), 113–120. https://www.cabdirect.org/cabdirect/abstract/20183366934.

Laperche, V., Traina, S. J., Gaddam, P., & Logan, T. J. (1996). Chemical and mineralogical characterizations of Pb in a contaminated soil: Reactions with synthetic apatite. *Environmental Science & Technology*, 30(11), 3321–3326. https://doi.org/10.1021/es960141u.

Lasat, M. (1999). Phytoextraction of metals from contaminated soil: A review of plant/soil/metal interaction and assessment of pertinent agronomic issues. *Journal of Hazardous Substance Research*, 2(1). https://doi.org/10.4148/1090-7025.1015.

Lewandowski, I., Clifton-Brown, J. C., Scurlock, J. M. O., & Huisman, W. (2000). Miscanthus: European experience with a novel energy crop. *Biomass and Bioenergy*, 19(4), 209–227. https://doi.org/10.1016/S0961-9534(00)00032-5.

Lewandowski, I., Clifton-Brown, J., Trindade, L. M., van der Linden, G. C., Schwarz, K.-U., Müller-Sämann, K., Anisimov, A., Chen, C.-L., Dolstra, O., Donnison, I. S., Farrar, K., Fonteyne, S., Harding, G., Hastings, A., Huxley, L. M., Iqbal, Y., Khokhlov, N., Kiesel, A., Lootens, P., … Kalinina, O. (2016). Progress on optimizing Miscanthus biomass production for the European bioeconomy: Results of the EU FP7 Project OPTIMISC. *Frontiers in Plant Science*, 7. https://doi.org/10.3389/fpls.2016.01620.

Lewandowski, I., Schmidt, U., & Faaij, A. P. C. (2005). Perennial crops are the most sustainable option for biomass production–A comparison of the efficiency and multiple land use options of perennial and annual biomass crops. *Industrial Ecology for a Sustainable Future (Abstract Book Oral Sessions)*, 144–146.

Li, Y.-M., Chaney, R., Brewer, E., Roseberg, R., Angle, J. S., Baker, A., Reeves, R., & Nelkin, J. (2003). Development of a technology for commercial phytoextraction of nickel: Economic and technical considerations. *Plant and Soil*, 249(1), 107–115. https://doi.org/10.1023/A:1022527330401.

Li, C., Wang, Q. H., Xiao, B., & Li, Y. F. (2011). *Phytoremediation potential of switchgrass (Panicum virgatum L.) for Cr-polluted soil*. 2011 International Symposium on Water Resource and Environmental Protection (ISWREP), Xi'an. https://doi.org/10.1109/ISWREP.2011.5893582.

López-Chuken, U. J., & Young, S. D. (2005). Plant screening of halophyte species for cadmium phytoremediation. *Zeitschrift für Naturforschung*, 60(3/4), 236–243.

Lotfy, S. M., & Mostafa, A. Z. (2014). Phytoremediation of contaminated soil with cobalt and chromium. *Journal of Geochemical Exploration*, 144, 367–373. https://doi.org/10.1016/j.gexplo.2013.07.003.

Lotfy, S. M., Mostafa, A. Z., & Abdel-Sabour, M. F. (2012). *Phytoextraction of cadmium from and zinc contaminated soils*. 3rd International Conference on Radiation Sciences and Applications, Hurghada, Egypt.

Ma, L. Q., Komar, K. M., Tu, C., Zhang, W., Cai, Y., & Kennelley, E. D. (2001). A fern that hyperaccumulates arsenic. *Nature*, 409(6820), 579–579. https://doi.org/10.1038/35054664.

Malayeri, B. E., Chehregani, A., Yousefi, N., & Lorestani, B. (2008). Identification of the hyper accumulator plants in copper and iron mine in Iran. *Pakistan Journal of Biological Sciences : PJBS*, 11(3), 490–492. https://doi.org/10.3923/pjbs.2008.490.492.

Mamirova, A., Pidlisnyuk, V., Amirbekov, A., Ševců, A., & Nurzhanova, A. (2020). Phytoremediation potential of *Miscanthus sinensis* And. In organochlorine pesticides contaminated soil amended by Tween 20 and Activated carbon. *Environmental Science and Pollution Research*. https://doi.org/10.1007/s11356-020-11609-y.

Matyka, M., & Kuś, J. (2016). Influence of soil quality for yielding and biometric features of *Miscanthus × giganteus*. *Polish Journal of Environmental Studies*, 25(1), 213–219. https://doi.org/10.15244/pjoes/60108.

Maughan, M., Bollero, G., Lee, D. K., Darmody, R., Bonos, S., Cortese, L., Murphy, J., Gaussoin, R., Sousek, M., Williams, D., Williams, L., Miguez, F., & Voigt, T. (2012). *Miscanthus × giganteus* productivity: The effects of management in different environments. *GCB Bioenergy*, 4(3), 253–265. https://doi.org/10.1111/j.1757-1707.2011.01144.x.

McCalmont, J. P., Hastings, A. F. S. J., McNamara, N. P., Richter, G. M., Robson, P., Donnison, I. S., & Clifton-Brown, J. C. (2015). Environmental costs and benefits of growing Miscanthus for bioenergy in the UK. *GCB Bioenergy*, 9, 489–507. https://doi.org/10.1111/gcbb.12294.

Medlin, E. A. (1997). An in vitro method for estimating the relative bioavailability of lead in humans [Master's thesis. Department of Geological Sciences, University of Colorado, Boulder, CO, USA].

Mi, J., Liu, W., Zhao, X., Kang, L., Lin, C., Yan, J., & Sang, T. (2018). N_2O and CH_4 emission from Miscanthus energy crop fields in the infertile Loess Plateau of China. *Biotechnology for Biofuels*, 11(1), 321. https://doi.org/10.1186/s13068-018-1320-8.

Miguez, F. E., Villamil, M. B., Long, S. P., & Bollero, G. A. (2008). Meta-analysis of the effects of management factors on *Miscanthus × giganteus* growth and biomass production. *Agricultural and Forest Meteorology*, 148(8), 1280–1292. https://doi.org/10.1016/j.agrformet.2008.03.010.

Moser, L., & Vogel, K. (1995). Switchgrass, big bluestem, and indiangrass. Publications from USDA-ARS/UNL Faculty. https://digitalcommons.unl.edu/usdaarsfacpub/2098.

Murányi, A., & Ködöböcz, L. (2008). *Heavy metal uptake by plants in different phytoremediation treatments.* VII Alps-Adria Scientific Workshop, Stara Lesna, Slovakia, 36, 387–390.

Narayanan, M., Erickson, L. E., & Davis, L. C. (1999). Simple plant-based design strategies for volatile organic pollutants. *Environmental Progress*, 18(4), 231–242. https://doi.org/10.1002/ep.670180409.

Nsanganwimana, F., Pourrut, B., Waterlot, C., Louvel, B., Bidar, G., Labidi, S., Fontaine, J., Muchembled, J., Lounès-Hadj Sahraoui, A., Fourrier, H., & Douay, F. (2015). Metal accumulation and shoot yield of *Miscanthus × giganteus* growing in contaminated agricultural soils: Insights into agronomic practices. *Agriculture, Ecosystems & Environment*, 213, 61–71. https://doi.org/10.1016/j.agee.2015.07.023.

Nurzhanova, A., Pidlisnyuk, V., Abit, K., Nurzhanov, C., Kenessov, B., Stefanovska, T., & Erickson, L. (2019). Comparative assessment of using *Miscanthus × giganteus* for remediation of soils contaminated by heavy metals: A case of military and mining sites. *Environmental Science and Pollution Research*, 26(13), 13320–13333. https://doi.org/10.1007/s11356-019-04707-z.

Ociepa, E. (2011). The effect of fertilization on yielding and heavy metals uptake by maize and Virginia fanpetals *(Sida hermaphrodita). Archives of Environmental Protection*, 37(2), 123–129.

Ogunkunle, C. O., Fatoba, P. O., Oyedeje, A. O., & Awotoye, O. O. (2014). Assessing the heavy metal transfer and translocation by *Sida acuta* and *Pennisetum purpureum* for phytoremediation purposes. *Albanian Journal of Agricultural Science*, 13(1), 71–80.

Pandey, V. C., & Bajpai, O. (2019). Chapter 1-Phytoremediation: From theory toward practice. In V. C. Pandey & K. Bauddh (Eds.), *Phytomanagement of Polluted Sites* (pp. 1–49), Elsevier, New York. https://doi.org/10.1016/B978-0-12-813912-7.00001-6.

Papazoglou, E. G., Karantounias, G. A., Vemmos, S. N., & Bouranis, D. L. (2005). Photosynthesis and growth responses of giant reed (*Arundo donax* L.) to the heavy metals Cd and Ni. *Environment International*, 31(2), 243–249. https://doi.org/10.1016/j.envint.2004.09.022.

Papazoglou, E. G., Serelis, K. G., & Bouranis, D. L. (2007). Impact of high cadmium and nickel soil concentration on selected physiological parameters of *Arundo donax* L. *European Journal of Soil Biology*, 43(4), 207–215. https://doi.org/10.1016/j.ejsobi.2007.02.003.

Parrish, D. J., Casler, M. D., & Monti, A. (2012). The evolution of switchgrass as an energy crop. In A. Monti (Ed.), *Switchgrass: A Valuable Biomass Crop for Energy* (pp. 1–28), Springer, London. https://doi.org/10.1007/978-1-4471-2903-5_1.

Pavel, P.-B., Puschenreiter, M., Wenzel, W. W., Diacu, E., & Barbu, C. H. (2014). Aided phytostabilization using *Miscanthus sinensis × giganteus* on heavy metal-contaminated soils. *Science of the Total Environment*, 479–480, 125–131. https://doi.org/10.1016/j.scitotenv.2014.01.097.

Pidlisnyuk, V., Mamirova, A., Pranaw, K., Shapoval, P. Y., Trögl, J., & Nurzhanova, A. (2020a). Potential role of plant growth-promoting bacteria in *Miscanthus × giganteus* phytotechnology applied to the trace elements contaminated soils. *International Biodeterioration & Biodegradation*, 155, 105103. https://doi.org/10.1016/j.ibiod.2020.105103.

Pidlisnyuk, V., Stefanovska, T., Lewis, E. E., Erickson, L. E., & Davis, L. C. (2014). Miscanthus as a productive biofuel crop for phytoremediation. *Critical Reviews in Plant Sciences*, 33(1), 1–19. https://doi.org/10.1080/07352689.2014.847616.

Pidlisnyuk, V. V., Erickson, L., Stefanovska, T., Popelka, J., Hettiarachchi, G., Davis, L., & Trögl, J. (2019). Potential phytomanagement of military polluted sites and biomass production using biofuel crop *Miscanthus × giganteus*. *Environmental Pollution*, 249, 330–337. https://doi.org/10.1016/j.envpol.2019.03.018.

Pidlisnyuk, V. V., Erickson, L. E., Trögl, J., Shapoval, P. Y., Popelka, J., Davis, L. C., Stefanovska, T. R., & Hettiarachchi, G. M. (2018). Metals uptake behaviour in *Miscanthus × giganteus* plant during growth at the contaminated soil from the military site in Sliač, Slovakia. *Polish Journal of Chemical Technology*, 20(2), 1–7. https://doi.org/10.2478/pjct-2018-0016.

Pidlisnyuk, V. V., Shapoval, P., Zgorelec, Z., Stefanovska, T., & Zhukov, O. (2020b). Multiyear phytoremediation and dynamic of foliar metal(loid)s concentration during application of *Miscanthus × giganteus* Greef et Deu to polluted soil from Bakar, Croatia. *Environmental Science and Pollution Research*, 27, 31446–31457. https://doi.org/10.1007/s11356-020-09344-5.

Pilu, R., Manca, A., & Landoni, M. (2013). Arundo donax as an energy crop: Pros and cons of the utilization of this perennial plant. *Maydica*, 58(1), 54–59. https://journals-crea.4science.it/index.php/maydica/article/view/916.

Pivetz, B. E. (2001). *Phytoremediation of contaminated soil and ground water at hazardous waste sites*. US Environmental Protection Agency, Office of Research and Development.

Pogrzeba, M., Krzyżak, J., & Sas-Nowosielska, A. (2013). Environmental hazards related to Miscanthus × giganteus cultivation on heavy metal contaminated soil. *E3S Web of Conferences*, 1, 29006. https://doi.org/10.1051/e3sconf/20130129006.

Pranaw, K., Pidlisnyuk, V., Trögl, J., & Malinská, H. (2020). Bioprospecting of a novel plant growth-promoting bacterium *Bacillus altitudinis* KP-14 for enhancing *Miscanthus × giganteus* growth in metals contaminated soil. *Biology*, 9(9), 305. https://doi.org/10.3390/biology9090305.

Pulford, I. D., & Watson, C. (2003). Phytoremediation of heavy metal-contaminated land by trees—A review. *Environment International*, 29(4), 529–540. https://doi.org/10.1016/S0160-4120(02)00152-6.

Rascio, N., & Navari-Izzo, F. (2011). Heavy metal hyperaccumulating plants: How and why do they do it? And what makes them so interesting? *Plant Science*, 180(2), 169–181. https://doi.org/10.1016/j.plantsci.2010.08.016.

Rinehart, L. (2006). *Switchgrass as a bioenergy crop*. National Center for Appropriate Technology, http://Attra.Ncat.Org/Attra-Pub/PDF/Switchgrass.Pdf.

Robinson, B. H., Bañuelos, G., Conesa, H. M., Evangelou, M. W. H., & Schulin, R. (2009). The phytomanagement of trace elements in soil. *Critical Reviews in Plant Sciences*, 28(4), 240–266. https://doi.org/10.1080/07352680903035424.

Roik, M., Sinchenko, V., Purkin, V., Kvak, V., & Humentik, M. (Eds.). (2019). *Miscanthus in Ukraine*. FOP Yamchinskiy Press.

Ruby, M. V., Davis, A., Schoof, R., Eberle, S., & Sellstone, C. M. (1996). Estimation of lead and arsenic bioavailability using a physiologically based extraction test. *Environmental Science & Technology*, 30(2), 422–430. https://doi.org/10.1021/es950057z.

Sabeen, M., Mahmood, Q., Irshad, M., Fareed, I., Khan, A., Ullah, F., Hussain, J., Hayat, Y., & Tabassum, S. (2013). Cadmium phytoremediation by *Arundo donax* L. from contaminated soil and water. *BioMed Research International*, 2013. https://doi.org/10.1155/2013/324830.

Sagehashi, M., Liu, C., Fujii, T., Fujita, H., Sakai, Y., Hu, H.-Y., & Sakoda, A. (2011). Cadmium removal by the hydroponic culture of giant reed (*Arundo donax*) and its concentration in the plant. *Journal of Water and Environment Technology*, 9(2), 121–127. https://doi.org/10.2965/jwet.2011.121.

Sankaran, R. P., & Ebbs, S. D. (2007). Cadmium accumulation in deer tongue grass (*Panicum clandestinum* L.) and potential for trophic transfer to microtine rodents. *Environmental Pollution*, 148(2), 580–589. https://doi.org/10.1016/j.envpol.2006.11.025.

Schmidt, C. S., Mrnka, L., Frantík, T., Lovecká, P., & Vosátka, M. (2018). Plant growth promotion of *Miscanthus ×giganteus* by endophytic bacteria and fungi on non-polluted and polluted soils. *World Journal of Microbiology and Biotechnology*, 34(3), 48. https://doi.org/10.1007/s11274-018-2426-7.

Shahandeh, H., & Hossner, L. (2000). Plant screening for chromium phytoremediation. *International Journal of Phytoremediation*, 2(1), 31–51. https://doi.org/10.1080/15226510008500029.

Shim, J., Babu, A. G., Velmurugan, P., Shea, P. J., & Oh, B.-T. (2014). *Pseudomonas fluorescens* JH 70-4 promotes Pb stabilization and early seedling growth of Sudan grass in contaminated mining site soil. *Environmental Technology*, 35(20), 2589–2596. https://doi.org/10.1080/09593330.2014.913691.

Sidella, S. (2014). *Adaptability, biomass yield, and phytoremediation of Arundo donax L. on marginal lands: Salt, dry and lead-contaminated soils* [Doctoral Thesis, University of Catania]. http://dspace.unict.it:8080/handle/10761/1605.

Solís-Dominguez, F. A., White, S. A., Hutter, T. B., Amistadi, M. K., Root, R. A., Chorover, J., & Maier, R. M. (2012). Response of key soil parameters during compost-assisted phytostabilization in extremely acidic tailings: Effect of plant species. *Environmental Science & Technology*, 46(2), 1019–1027. https://doi.org/10.1021/es202846n.

Tari, I., Poór, P., Ördög, A., Székely, A., Laskay, G., & Bagi, I. (2013). Enhanced biomass production in sudangrass induced by co-treatment with copper and EDTA. *Environmental and Experimental Biology*, 11, 151–157.

Teel, A. (2003). Management guide for the production of switchgrass for biomass fuel in southern Iowa. PM 1710, May 2003, 3 pages. https://www.agmrc.org/media/cms/PM1710_097EE99B56173.pdf

Tordoff, G. M., Baker, A. J. M., & Willis, A. J. (2000). Current approaches to the revegetation and reclamation of metalliferous mine wastes. *Chemosphere*, 41(1), 219–228. https://doi.org/10.1016/S0045-6535(99)00414-2.

Tryboi, O. V. (2018). Efficient biomass value chains for heat production from energy crops in Ukraine. *Energetika*, 64(2), Article 2. https://doi.org/10.6001/energetika.v64i2.3782.

USEPA. (1998). *A citizen's guide to phytoremediation* (EPA 542-F-98-011). U.S. Environmental Protection Agency, Office of Solid Waste and Emergency Response.

USEPA. (2000). *Introduction to phytoremediation* (EPA 600/R-99/107). U.S. Environmental Protection Agency.

Vassey, T. L., George, J. R., & Mullen, R. E. (1985). Early-, mid-, and late-spring establishment of switchgrass at several seeding rates. *Agronomy Journal*, 77(2), 253–257. https://doi.org/10.2134/agronj1985.00021962007700020018x.

Vogel, K. P. (1987). Seeding rates for establishing big bluestem and switchgrass with preemergence atrazine applications. *Agronomy Journal*, 79(3), 509–512. https://doi.org/10.2134/agronj1987.00021962007900030021x.

Vogel, K. P. (2000). Improving warm-season forage grasses using selection, breeding, and biotechnology. In Kenneth J. Moore & Bruce E. Anderson (Eds.), *Native Warm-Season Grasses: Research Trends and Issues* (pp. 83–106). John Wiley & Sons, Ltd. https://doi.org/10.2135/cssaspecpub30.c6.

Wang, C., Kong, Y., Hu, R., & Zhou, G. (2020). Miscanthus: A fast-growing crop for environmental remediation and biofuel production. *GCB Bioenergy*, 13, 58–69. https://doi.org/10.1111/gcbb.12761.

Wijesekara, H., Bolan, N. S., Vithanage, M., Xu, Y., Mandal, S., Brown, S. L., Hettiarachchi, G. M., Pierzynski, G. M., Huang, L., Ok, Y. S., Kirkham, M. B., Saint, C. P., & Surapaneni, A. (2016). Chapter Two - Utilization of biowaste for mine spoil rehabilitation. In D. L. Sparks (Ed.), *Advances in Agronomy* (Vol. 138, pp. 97–173). Academic Press, Cambridge, MA. https://doi.org/10.1016/bs.agron.2016.03.001.

Wilkins, C. (1997). The uptake of copper, arsenic and zinc by Miscanthus—Environmental implications for use as an energy crop. *Aspects of Applied Biology*, No. 49, 335–340. https://www.cabdirect.org/cabdirect/abstract/19970707775.

Wolf, D. D., & Fiske, D. A. (2009). *Planting and managing switchgrass for forage, wildlife, and conservation.* https://vtechworks.lib.vt.edu/handle/10919/50258.

Xia, H. P. (2004). Ecological rehabilitation and phytoremediation with four grasses in oil shale mined land. *Chemosphere*, 54(3), 345–353. https://doi.org/10.1016/S0045-6535(03)00763-X.

Yang, M., Xiao, X., Miao, X., Guo, Z., & Wang, F. (2012). Effect of amendments on growth and metal uptake of giant reed (*Arundo donax* L.) grown on soil contaminated by arsenic, cadmium and lead. *Transactions of Nonferrous Metals Society of China*, 22(6), 1462–1469. https://doi.org/10.1016/S1003-6326(11)61342-3.

Yang, S., Liao, B., Li, J., Guo, T., & Shu, W. (2010). Acidification, heavy metal mobility and nutrient accumulation in the soil–plant system of a revegetated acid mine wasteland. *Chemosphere*, 80(8), 852–859. https://doi.org/10.1016/j.chemosphere.2010.05.055.

Zema, D. A., Bombino, G., Andiloro, S., & Zimbone, S. M. (2012). Irrigation of energy crops with urban wastewater: Effects on biomass yields, soils and heating values. *Agricultural Water Management*, 115, 55–65. https://doi.org/10.1016/j.agwat.2012.08.009.

Zgorelec, Ž. (2009). *Phytoaccumulation of metals and metalloids from soil polluted by coal ash* [University of Zagreb]. https://www.bib.irb.hr/439719?rad=439719.

Zgorelec, Z., Bilandzija, N., Knez, K., Galic, M., & Zuzul, S. (2020). Cadmium and mercury phytostabilization from soil using *Miscanthus* ×*giganteus*. *Scientific Reports*, 10(1), 6685. https://doi.org/10.1038/s41598-020-63488-5.

Zhang, J., Yang, S., Huang, Y., & Zhou, S. (2015). The tolerance and accumulation of *Miscanthus sacchariflorus* (maxim.) Benth., an energy plant species, to cadmium. *International Journal of Phytoremediation*, 17(6), 538–545. https://doi.org/10.1080/15226514.2014.922925.

Zhang, X., Xia, H., Li, Z., Zhuang, P., & Gao, B. (2010). Potential of four forage grasses in remediation of Cd and Zn contaminated soils. *Bioresource Technology*, 101(6), 2063–2066. https://doi.org/10.1016/j.biortech.2009.11.065.

Zhang, Y., Xu, C., Lu, J., Yu, H., & Xia, T. (2020). An effective strategy for dual enhancements on bioethanol production and trace metal removal using Miscanthus straws. *Industrial Crops and Products*, 152, 112393. https://doi.org/10.1016/j.indcrop.2020.112393.

Zhuang, P., Shu, W., Li, Z., Liao, B., Li, J., & Shao, J. (2009). Removal of metals by sorghum plants from contaminated land. *Journal of Environmental Sciences*, 21(10), 1432–1437. https://doi.org/10.1016/S1001-0742(08)62436-5.

Żurek, G., Pogrzeba, M., Rybka, K., & Prokopiuk, K. (2013). Suitability of grass species for phytoremediation of soils polluted with heavy-metals. In S. Barth & D. Milbourne (Eds.), *Breeding Strategies for Sustainable Forage and Turf Grass Improvement* (pp. 245–248), Springer, Dordrecht.

Zwonitzer, J. C., Pierzynski, G. M., & Hettiarachchi, G. M. (2003). Effects of phosphorus additions on lead, cadmium, and zinc bioavailabilities in a metal-contaminated soil. *Water, Air, & Soil Pollution*, 143, 193–209. https://doi.org/10.1023/A:1022810310181.

3

Remediation of Sites Contaminated by Organic Compounds

Lawrence C. Davis, Barbara Zeeb, Larry E. Erickson,
Aigerim Mamirova, and Valentina Pidlisnyuk

Abstract

The transformation and biodegradation of organic contaminants in soils with plants occur in plants as well as in soil. Microorganisms have the ability to biodegrade many compounds and microbial populations are larger when plants are present because of root exudates. In this chapter, petroleum compounds, explosives, solvents, pesticides, and persistent organic pollutants are included. Miscanthus, trees, and many other plants have important phytoremediation applications to organic contaminants. Phytoremediation studies with Miscanthus show that tolerance to organic contaminants is good and that Miscanthus is an effective plant for phytoremediation. Positive biodegradation results are reported with hemp, which is another plant with commercial value. Because of the importance to NATO, phytoremediation research progress with explosives is an important part of this chapter. There has been good research progress in phytoremediation applications with poplar trees where polychlorinated biphenyls have been investigated. Some recent phytoremediation advances with dioxins are included. Recent phytoremediation results with Miscanthus growing in pesticide-contaminated soil show that Miscanthus is able to grow in soils where mixtures of chlorinated pesticides are present.

CONTENTS

3.1 Introduction

There have been many studies of phytoremediation where organic contaminants are present and are the focus of the research or the cleanup. Phytoremediation with biomass production to obtain useful products is the emphasis in this book, but is far less common. If remediation goals at a site can be profitable because a useful product is harvested and sold, this has value for the project. With organic contaminants, it is possible to restore land to a state of productive use. In this chapter the emphasis will be on both soil remediation and how it can be accomplished by plants that have economic value. *Miscanthus ×giganteus* (*M. ×giganteus*) is the primary focus of this book, but because there is a dearth of information about the use of that plant in remediation of organics, most examples are drawn from other species.

3.2 Types of Organic Contaminants

There are many organic compounds present in soil and ground water as contaminants. Petroleum hydrocarbon (PHC) contaminated soil is one of the major areas of investigation and application as PHCs are among the most prevalent pollutants in the environment (e.g., Abdullah et al., 2020). For example, in Canada, approximately 60% of contaminated sites involve PHC contamination, often impairing the quality and uses of land and water (CCME, 2008). Hence, large land areas such as closed petroleum refineries, soils associated with restoration of coalmine lands, and spills at petroleum production areas need to be remediated. Explosives in soil are important in this book because of the emphasis on content that is of interest to NATO. Wood treatment sites with creosote, aircraft de-icing chemicals near airports, and many types of solvents at dry cleaners and vehicle repair shops are present in many countries. Persistent organic pollutants (POPs) such as the industrial chemicals, polychlorinated biphenyls (PCBs), and pesticides like DDT are present in soil as contaminants at many locations worldwide (Tarla et al., 2020). These organic contaminants adversely affect human and environmental health globally as they are subject to long-range transport via slow global distillation and persist in soils long after their initial deployment (Chlebek & Hupert-Kocurek, 2019).

3.2.1 Remediation of Petroleum Contaminants

Petroleum spills and leaking tanks have resulted in many sites where organic compounds are present in soil. There is a significant literature on phytoremediation and bioremediation of soils with petroleum contaminants

(Chan-Quijano et al., 2020; Fiorenza et al., 2000; McCutcheon & Schnoor, 2003; Tang, 2019). Locations with petroleum and natural gas production operations, pipelines, refineries, storage sites, gasoline stations, vehicle maintenance shops, and parking lots are some of the places where hydrocarbons are routinely found in soils. Leaking underground storage tanks have been found at many locations including service stations, buildings where heating oil is used, and on farms where fuel for vehicles is stored.

Most PHCs have very low solubility in water and often the separation of a two-phase mixture of oil and water allows some valuable recovery of the oil phase when a "pump and treatment" system is used to recover product. Gasoline, kerosene, and diesel fractions are liquid at ambient temperature; however, some petroleum compounds are solids under ambient temperatures. When microorganisms feed on petroleum compounds, they may be found at oil phase surfaces. Polycyclic aromatic hydrocarbons (PAHs) are natural petrogenic materials although they can also be products of incomplete combustion of hydrocarbon fuels. Often, they are the most hazardous components of fuel spills because many PAHs are classified as carcinogenic (Cachada et al., 2016; Henner et al., 1997). Loss from soil to atmosphere is an important route of dispersal for petroleum fractions with reasonable vapor pressures, and fairly low water solubility. For instance, gasoline, kerosene, and jet fuel will dissipate from soils relatively quickly if the soil is porous. Plants which remove water from soils often facilitate diffusive loss of contaminants trapped beneath the water table by lowering the water table.

Petroleum compounds provide carbon and energy, but nitrogen (N) and phosphorus (P) are also needed to support growth of both the plants and microorganisms. Because N and P are needed, fertilizer is often added as part of the phytoremediation plan. Organic fertilizers such as manure add to microbial diversity, which is often beneficial (Chan-Quijano et al., 2020). Many different bacteria participate in the biodegradation of PHCs, and some authors have generated lists of different species that have been isolated (e.g., Chan-Quijano et al., 2020). Research on the biodegradation of hydrocarbons was ongoing when phytoremediation research to address PHC-contaminated soil began in the 1990s.

Several of the early field studies in the 1990s were carried out by Banks and coworkers (Fiorenza et al., 2000). The results from several field studies are included in McCutcheon & Schnoor (2003). Grasses have been among the more beneficial plants in the research on biodegradation of petroleum compounds (Chapter 11 by Hutchinson, Banks and Schwab in McCutcheon & Schnoor (2003)). Although many forage grasses have only very limited use in industrial biomass production, the early research demonstrated beneficial effects of plants in field-scale PHC degradation. Plant roots provide an active environment, which contains compounds and organisms that are beneficial for microbial degradation processes. Often, they enhance soil aeration by removing pore water. Many degradative reactions are oxygen-dependent, although some are anaerobic.

One good example of phytoremediation using trees for removing fuel contaminants from a shallow aquifer was reported by Nichols et al. (2014). In their study, ~579,000 L of diesel, jet fuel, and gasoline (all moderately volatile) were present at the start of the project in an area of two hectares. Poplar, willow, and pine trees were planted at the site in 2006 with most of the 3250 trees being poplars. When poplar and willow trees died, they were replaced by cuttings from the healthy trees at the site. Soil-gas sampling was used to follow the progress of the remediation and determined a 95% loss of total PHCs (TPH), and a 99% loss of mass of benzene (very volatile). As the trees grew, their ability to pump water increased and this was beneficial. In this example, the TPHs were all liquids at ambient conditions. Methyl-tert-butyl ether (MTBE), a highly water-soluble fuel additive, was taken up by the trees and released to the atmosphere. As the rate of release to the atmosphere was limited by the rate of evapotranspiration of the water that the MTBE was dissolved in, the concentration of MTBE in the atmosphere was very small (Narayanan et al., 1999), and in addition, MTBE has a very short atmospheric half-life on the order of 3 days (Squillace et al., 1997). There is no doubt that other constituents passed through the trees at lower levels proportional to their water vs lipid solubility. Toxicity to plants would limit this uptake for levels of benzene, toluene, ethylbenzene, and xylenes. Less polar lipids, found in crude oil and environmentally aged petroleum fractions, are generally much less toxic.

Combining the use of vascular plants and microbes (bacteria and/or fungi) is proving to be a promising approach for degrading a variety of organic contaminants including PHCs. Studies carried out recently have largely focused on the potential of endophyte and rhizosphere plant growth promoting bacteria to increase the efficiency of phytodegradation (e.g., Becerra-Castro et al., 2013; Chlebek & Hupert-Kocurek, 2019). These bacteria, possessing catabolic genes, mineralize organic contaminants within the plant or rhizosphere, reducing their phytotoxicity, while promoting the growth and development of plant root and shoot biomass (e.g., Afzal et al., 2014; Arslan et al., 2017; Glick, 2010; Santoyo et al., 2016).

Cannabis sativa is an annual dioecious herb capable of growing to heights of 5 m and having long tap roots. This plant has been grown since ancient times for use in a wide range of applications. Fibers from hemp are extensively used in products that include fabrics and textiles, ropes, yarn, carpeting, construction and insulation materials, etc. (Johnson, 2014). The short and woody fibers in the hemp's stalk interior are known as "hurds" and are used in the manufacture of animal bedding, paper, and composites. Hemp seed is used in various foods and beverages and oil from hemp seed is widely used in industrial oils, cosmetics, and pharmaceuticals. In addition, cannabinoids, a group of compounds found in Cannabis (with the most notable being the phytocannabinoid tetrahydrocannabinol (THC)), are used medicinally, spiritually, and recreationally (e.g., Bilalis et al., 2019). The term "hemp" is used to denote cannabis that contains 0.3% or less THC content by dry weight. In recent years, *C. sativa* (hemp) has been studied as a bioenergy crop as it grows well on marginal lands and has the capacity to produce high volumes of biomass (e.g., Kumar et al., 2017). Asquer et al. (2019) reported biogas production using hemp straw was comparable to most other energy crops. This annual crop can be grown in climates where winters are too cold, making successful maintenance of perennial *M. ×giganteus* unreliable.

   C. sativa (hemp) was studied in the remediation of two PAHs, benzo[α] pyrene and chrysene (Campbell et al., 2002). The authors carried out experiments over 45 days in soil spiked with benzo[α]pyrene and chrysene at 25, 50, and 75 µg g^{-1} and found reductions in contaminants in all cases. They additionally found that the mass and growth of Cannabis plants to increase at all three concentrations leading them to suggest that metabolites of the two PAHs studied may have stimulated the growth of hemp.

   Research to improve PHC phytoremediation processes is continuing, and there has been significant progress in the last 6 years. In 2019, Tang reviewed studies on the biodegradation of TPHs (Tang, 2019). Ren et al. (2017) discuss in detail some of the complexity in remediating PHC because the larger aromatics are tightly sorbed to organic matter, and minerals of soil giving them very low accessibility to organisms. The sorption/desorption processes are challenging to predict for the vast number of constituents in PHC mixtures such as crude oil, and residuals from refining. Laboratory experimentation may not align well with effects observed in the field with "aged" materials because of the very slow rates of sorption/desorption observed in real soil structures, particularly in micropores of mineral or biochar fractions (Ren et al., 2017).

3.2.2 Remediation of Explosives

Large areas of land are contaminated with explosives or their residues. Landmines, unexploded ordnance (UXO), and explosive compounds in soil are important issues in many countries. Globally more than 80 countries have land contaminated by explosives (Robledo et al., 2009), including more than 100 million antipersonnel mines (Hemapala, 2017). In Europe, there are UXOs from World War 2 that still need to be removed. For example, Ukraine has ~7000 km^2 of land with UXOs, and this resulted in 2078 casualties from 2014 to 2017 (Dathan, 2020). Due to military conflicts in Asia and the Middle East, more than 150 million ha have explosives present as UXOs and/or as contaminants in soil (Via, 2020). In the USA, there are more than 2000 sites with soil contamination due to explosives (Via, 2020). Contaminated soil locations include sites for the manufacture of explosives, assembly plants where explosives are, or were, packed into shells, and sites where explosives have been stored.

   There has been good progress in developing robotic methods to identify and remove landmines, one of the principal forms of UXO (Hemapala, 2017; Robledo et al., 2009). Kalderis et al. (2011) provided a comprehensive review of research on the biodegradation of TNT (2,4,6-trinitrotoluene), RDX (Royal Demolition Explosive; cyclotrimethylene-trinitramine; 1,3,5-trinitroperhydro-1,3,5-triazine), and HMX (high-melting explosive; octogen; cyclotetramethylene-tetranitramene; 1,3,5,7-tetranitro-1,3,5,7-tetrazocane). Water solubility is largest for TNT (130 mg L^{-1}), followed by RDX (42 mg L^{-1}), and HMX (5 mg L^{-1}) facilitating their transfer to organisms. Biodegradation pathways are presented in the Kalderis' et al. (2011) review. The greatest success to date with bioremediation and phytoremediation has been with TNT, the simplest and most soluble of the common explosives. Lists of bacteria and fungi that biodegrade TNT, RDX, and HMX are provided in the Kalderis et al.'s (2011) review along with data on toxicity of these explosive compounds to microorganisms and invertebrates.

Phytoremediation field studies of TNT have been reported, and a good summary of the few reported, is provided by Via (2020), along with degradation pathways of TNT within plants, and a list of plants with results. A field-scale wetland treatment system for TNT and RDX in water was designed and implemented at the Iowa Army Ammunition Plant (McCutcheon & Schnoor, 2003).

The concept of phytoremediation with biomass production on sites with explosive contaminants needs further development. One tropical biomass crop, vetiver grass (*Chrysopogon zizanoides*), a relative of sorghum, has been tested for its capacity to take up and degrade TNT (Das, 2014). When urea was added to the soil system, uptake increased as much as 90% of the input 100 mg kg^{-1} amount of TNT in 22 days. Miscanthus may be a good plant for field-scale phytoremediation of soils contaminated with explosive compounds; however, further research is needed on the fate of TNT, RDX, and HMX in soils where Miscanthus is grown. In many cases, the actual degradation is microbially driven, and hence many different plant species may facilitate the process (Esteve-Núñez et al., 2001). On the other hand, TNT metabolism and detoxification within plants may differ between plant species. The process has been well characterized in the dicot *Arabidopsis* (Gandia-Herrero et al., 2008), but thus far not in grasses.

3.2.3 Remediation of Chlorinated Hydrocarbons

Many chlorinated organic compounds have found their way into soil and ground water. The liquid forms are especially challenging to deal with because they are often denser than water and move to the bottom of an aquifer, gradually contaminating that water by slow diffusive dissolution. Chlorinated solvents such as trichloroethylene have been used for a number of beneficial applications in machine shops and dry-cleaning operations, and they are found in soil and groundwater at many locations. Volatile solvents dissolved in water can be taken up into plant roots. Plant evapotranspiration releases water into the atmosphere and the chlorinated solvents in the water are released into the atmosphere. Because of the low vapor pressure of water at ambient temperatures, only about 18 mg L^{-1} of water (1 mM) can be evaporated into the air phase. In consequence, huge volumes of air are needed for evapotranspiration, and thus the concentrations of the volatile chlorinated solvents in the air are very low. The dilution factor of this process is about 55,000, varying with temperature and relative humidity of the "incoming" air (Narayanan et al., 1995, 1999). In the atmosphere, trichloroethylene has a short half-life as solar radiation generates hydroxyl radicals.

Keeping soluble chlorinated organic compounds out of drinking water sources is challenging. Once they have found their way into an aquifer, *in situ* treatment in the aquifer may be a better treatment alternative than pumping the water out (Santharam et al., 2011). The effectiveness of phytoremediation is limited by the ability of roots to reach the contamination. Sometimes efforts are made to install trees below the soil surface, in an excavated pit or well, in order to reach the contaminated zone (Negri et al. in McCutcheon & Schnoor, 2003).

There are many examples of phytoremediation field studies where microbial transformations and evapotranspiration of chlorinated aliphatic compounds are observed (McCutcheon & Schnoor, 2003). The vegetation effectiveness is better for contaminants that are near the soil surface compared to aquifers that are located deeper below the ground surface. Vegetation has been used to pump contaminated water into the atmosphere as a way to manage contaminants that are present at a site (Doucette et al. in McCutcheon & Schnoor, 2003). An alternative is to pump contaminated water to the surface and use it to irrigate trees or other vegetation (Jordahl et al. in McCutcheon & Schnoor, 2003). Deep-rooted, water-seeking trees have been used effectively and economically for remediation of chlorinated compounds (Shang et al. in McCutcheon & Schnoor, 2003). They have value when harvested.

PCBs are classic examples of POPs. They are primarily an intentional industrial product, used for 50 years or more as an insulating oil (liquid) in capacitors and electrical transformers (USEPA, n.d.). Their extreme hydrophobicity results in strong sorption to organic matter in soils. As discussed by Ren et al. (2017), the nature of sorption may vary with the nature of the soil matrix, depending on the specific types of transformed organic matter present, and types of clay and other minerals in the soil. Thus, different PCB congeners may sorb/desorb differently on different soils. The same caveat applies to dioxins and pesticides (see below).

The ability of microbe and plant species to access these POPs varies greatly, depending on whether the plant or microbe produces surfactants, or specific lipid binding proteins in the rhizosphere (Terzaghi et al., 2021). Those authors compared seven species and some combinations of species, including pumpkin with fescue, and the effect of compost or redox cycling with fescue. Other species were treated with other methods including redox cycling, pairwise growth, or ammonium thiosulfate addition. Fescue and a combination with pumpkin gave the best rates of degradation of the complex mixture of PCBs, of which they analyzed 79 congeners. This included the most highly chlorinated classes. They noted that levels of organic C, as compost, altered rates. It was proposed that it serves as a source of dissolved organic carbon which facilitates microbial degradation of PCBs.

Biochar, the charcoal obtained by incomplete combustion of plant material, may alter the relative sorption of POPs. For example, in a greenhouse experiment, the addition of 2.8% (by weight) biochar to soil contaminated with 136 and 3.1 µg g^{-1} of PCBs, reduced PCB root concentration in the known phyto-extractor *Cucurbita pepo* ssp. *pepo* by 77% and 58%, respectively, in addition to increasing aboveground plant biomass (Denyes et al., 2012). Further, in the first *in situ* experiment conducted at a Canadian PCB-contaminated Brownfield site, two types of biochar were statistically equal at reducing PCB uptake into plants as granular activated carbon (AC), reducing PCB concentrations in *C. pepo* root tissue by up to 74% (Denyes et al., 2013). Biochar-equivalent may be a natural material in black soils, where the black color is due to ancient and modern products of fires. Nartey & Zhao (2014) thoroughly review various processes for the production of biochars and Denyes et al. (2014) discuss the importance of their physical, chemical, and biological characterization.

Ficko et al. (2011) conducted a field study in which three promising phyto-extracting perennial weed species (*Chrysanthemum leucanthemum*, *Rumex crispus*, and *Solidago canadensis*) were planted in monoculture plots at two PCB-contaminated sites in southern Ontario and followed over 2 years to investigate the effects of plant age, contaminant characteristics, and species-specific properties on PCB uptake and accumulation patterns in plant tissues. Results indicated that shoot contaminant concentrations and total biomass were dependent on plant age and life cycle (vegetative and reproductive stages), which affected the total amount of PCBs phyto-extracted on a per-plant basis. Even at suboptimal planting densities of 3–5 plants m^{-2}, all three weed species extracted a greater quantity of PCBs per unit area (4800–10,000 µg m^{-2}) than the known PCB-accumulator *Cucurbita pepo*. Calculated PCB extractions based on theoretical optimal planting densities were significantly higher at both sites and illustrated the potential of these weeds for site remediation.

An excellent example of plant-assisted remediation using trees may be found in Ancona et al. (2017). They used poplar trees, with drip irrigation, to remediate a site near transformers at a power station in southern Italy. At this site, which was also used as a dump for assorted wastes, the long-term spillage of PCB oils followed by recent efforts to clean up the site had dispersed into the soil to depths up to 40 cm. Within 1 year of planting trees, in rows spaced 2 m apart with trees at 0.5 m within the rows, levels of many congeners decreased from more than five-fold above regulatory limits to levels below those limits. The effect decreased as distance from the trees increased. Some lesser chlorinated congeners were taken up into the trees in limited amounts, while other more hydrophobic (more chlorinated) ones sorbed tightly to the roots. Overall, soil levels decreased >90%, and levels within the trees did not exceed those of the rhizosphere, despite large uptake of water over the course of a year. This result is not unexpected. Ancona et al. (2017) cite more than 50 articles describing microbial and plant-assisted degradation of PCBs, though mostly in pot studies. Chekol et al. (2004) documented the rhizosphere effect for PCBs with three legumes and four grasses, while Liu & Schnoor (2008) quantitatively documented the preferential uptake of lesser chlorinated PCBs into hydroponic poplars.

Dioxins (polychlorodibenzo-*p*-dioxins) are chlorinated aromatic compounds similar to PCBs but having two oxygen bridges between two benzene rings. Polychlorinated benzofurans have one oxygen bridge and one direct benzene-benzene bond. They arise as by-products of chlorination of phenols and during combustion of organic matter in the presence of O and Cl (Campanella et al., 2002). All of these are toxic to greater or lesser extent, but very persistent in environment, and so nonpolar that they accumulate in lipids of living organisms, and hence through the food chain. Fungi and root-associated organisms may facilitate their degradation. Bacteria of several genera are known to degrade specific congeners with varying efficiency. Many are aerobes, but some dechlorinations occur with anaerobes (Field & Sierra-Alvarez, 2008). Plants have been used to enhance the rates of degradation by providing nutrients to microbes, aerating the root zone, inducing microbial metabolism enzymes, and to do some metabolic reactions within the plant (Campanella et al., 2002). Over the past two decades there have been advances in this area, but challenges remain (Mench et al., 2010). There are many more recent papers discussing degradation of these compounds by microbes (Saibu et al., 2020), but little new work with plant pathways.

3.2.4 Remediation of Pesticides

At many locations in the world, soils are contaminated with pesticides because of poor management, spills, and the need to clean out sprayers. A recent review describes applications of phytoremediation to restore lands that are contaminated with pesticides (Tarla et al., 2020). The toxicity of pesticides to vegetation is one of the challenges in selecting plants for use at a contaminated site. There has been progress in the identification and use of microorganisms and plants that have good capability to degrade some pesticides. Soil amendments such as manure or biochar can be added to contaminated sites to dilute pesticides, sorb them out of soil solution, and provide additional microorganisms to aid in the establishment of vegetation. Use of vegetation that naturally grows well in the region is recommended.

Khalid et al. (2020) review many applications of biochar specific to pesticides. Deliberate augmentation with biochar to enhance bioremediation is becoming a common practice; however, sorption to biochar also reduces availability of pesticides and herbicides to plants, and can thus delay their remediation (Khalid et al., 2020). For example, in an *in situ* study at Point Pelee National Park in Canada (PPNP), biochar significantly reduced DDT (p,p'-dichlorodiphenyl-trichloroethane) accumulation in earthworms (49%) but did not significantly reduce plant uptake of DDT (Denyes et al., 2016). Other amendments, particularly other carbon-rich materials including humic substances, charcoal, bio-coal and AC, sorb the contaminants in soils (Beesley et al., 2011) and may also reduce the bioavailability for plants (Khalid et al., 2020). It is important to note that as shown by Nartey & Zhao (2014) different kinds of biochar prepared in different ways from different starting materials may have different effects.

Among pesticides, there is a large class of obsolete pesticides (insecticides) containing significant amounts of chlorination, resulting in persistence. Examples include aldrin/dieldrin/endrin, lindane (α-hexachlorocyclohexane), chlordane, chlordecone (Kepone), DDT, heptachlor, toxaphene, and others. Many of these are banned in many countries, but residues and waste/abandoned sites are still present (Agbeve et al., 2013; Tarla et al., 2020). Weathered DDT is still found in many locations, and portions of Africa continue to use it for control of malaria-bearing mosquitos (Tarla et al., 2020). Many other pesticides are present in agricultural/horticultural soils and in plants harvested from such soils. For instance, Agbeve et al. (2013) found β-HCH, δ-HCH, γ-HCH, heptachlor, aldrin, γ-chlordane, α-endosulfan, p.p'-DDE (2,2'-p,p'dichlorodiphenyl-1,1-dichloroethene), dieldrin, endrin, β-endosulfan, p.p'-DDD (2,2'-p,p'dichlorodiphenyl-1,1-dichloroethane, p.p'-DDT, and methoxychlor in roots of *Cryptolepis sanguinolenta*, a traditional antimalarial herb of Ghana, in both dry and rainy season's growth. Roots were thoroughly washed before analysis, which may have reduced the pesticide load, though perhaps not completely, as sorption is quite strong on root surfaces. The detected levels were below regulatory limits except for some samples with high aldrin/dieldrin. It is not known whether the various pesticides were long-term residues, or from banned products still being used. Tarla et al. (2020) discuss some of the problems with waste pesticide disposal.

There is potential for phytoremediation of DDT because it is taken up by some vascular plants (Lunney et al., 2004) particularly some species of the *Cucurbitaceae* family. Zucchini, a popular edible vegetable, accumulates large amounts from contaminated soil. It also accumulates aldrin, dieldrin, and endrin, such that it may be used to remove these residues from agricultural lands long contaminated with high levels, above current regulatory limits (Otani et al., 2007). However, very few species take up these compounds (only 2 of 15 families tested by Otani et al. (2007)), limiting ability to clean up many sites.

Paul et al. (2015) conducted a field investigation at three DDT-contaminated areas in PPNP in Canada. *Cucurbita pepo* (pumpkin) and three native grass species, *Schizachyrium scoparium*, *Panicum virgatum*, and *Sporobolus cryptandrus*, were grown at three different sites in PPNP having low (291 ng g^{-1}), moderate (5083 ng g^{-1}), and high (10,192 ng g^{-1}) soil DDT contamination levels. A threshold soil DDT concentration was identified at ~5000 ng g^{-1} where the DDT uptake into *C. pepo* was maximized, resulting in plant shoot and root DDT concentrations of 16,600 and 45,000 ng g^{-1}, respectively. Two of the native grass species (*P. virgatum* and *S. scoparium*) were identified as potential phytoextractors, with higher shoot extraction capabilities than that of the known phytoextractor *C. pepo* when optimal planting density was taken into account. Hexachlorocyclohexane, a common persistent contaminant (Agbeve et al., 2013; Tarla et al., 2020), has been remediated using plants also (Becerra-Castro et al., 2013).

White and Kottler (2002) used citric acid to augment the remediation of weathered DDT/DDE (DDT metabolite) with four species of plants including clover, hairy vetch, mustard, and ryegrass. Uptake of DDE was significantly increased while the region around the roots was depleted in concentration. Other surfactants and organic acids have been used to enhance the release of POPs. This is commonly characterized as chemically augmented phytoremediation. In order to improve the phytoremediation potential of plants and microbes, one may try using surface-active compounds including synthetic or biosurfactants (Agnello et al., 2014; An et al., 2011; Ramamurthy & Memarian, 2012); several different organic acids (Agnello et al., 2014; Gonzalez et al., 2010); and nanoparticles (Pillai & Kottekottil, 2016; Rani et al., 2017).

Herbicides, fungicides, and insecticides produced and used in recent years tend to have less residual action and decreased environmental stability than highly chlorinated POPs. Nevertheless, their remediation/detoxification is still necessary to avoid toxic effects on nontarget organisms during accidental exposure, and from field run-off during sudden rains, as they are typically applied at concentrations that are >100 × toxic levels (e.g., Fairchild et al., 1998; Plhalova et al., 2012).

Loffredo et al. (2020) found *Cannabis sativa* L. seedlings to be very effective in removing the systemic fungicide metalaxyl-M from water. Residual compounds accumulated in the hemp tissues over 7 days were much lower than the amounts removed from the medium, suggesting efficient metabolization. When hemp was allowed to germinate and grow in columns filled with soil contaminated with metalaxyl-M and the herbicide, metribuzin, it showed a noticeable remediation capacity. The authors used this study to suggest that hemp is a promising candidate for phytoremediation of wastewater and soil from pesticides.

3.3 Landfills and Containment

Containment of contaminants is an issue at landfills where waste is deposited and managed. In order to control contaminant movement within landfills, trees are often planted on the down gradient side to remove contaminated leachate from the site. This concept has also been used at the edges of fields to capture agricultural chemicals with a riparian strip to prevent them from entering a creek or river. There are significant savings when vegetation is used for pump-and-treat applications (Negri et al. in McCutcheon & Schnoor, 2003). The organic contaminants may be biodegraded or transpired and destroyed in the atmosphere. Trees that have the ability to reach the water table and have value when harvested are generally recommended for these applications. Trees have been used to some extent directly on landfill covers, but native plants, including shrubs and forbs, are recommended (USEPA, 2015). Deep-rooted grasses may also be effective. Grasses have been widely used for landfill covers to reduce water infiltration, but less so as a means to control leachate, once it is formed. Gąbka & Wolski (2011) described a successful study of active management of leachate by watering various turf grasses with collected landfill leachate at a closed landfill in Poland. In warm subtropical to tropical climates, vetiver grass has been used since 1994 and shown to be highly effective for managing a wide range of leachates either actively by irrigation, or passively by planting directly into the leachate seepage path (Vetiver Network International, 2017). Miscanthus has suitable characteristics for temperate climates including deep rooting where potential crop evapotranspiration exceeds normal precipitation (see Chapter 5 for information on water usage by Miscanthus).

3.4 Phytoremediation of Organic Contaminants with Miscanthus

Miscanthus is a C_4 grass related to sugarcane and sorghum, with a rich microbiome. Thus, it is anticipated that it can facilitate rhizoremediation of many organic compounds. Field data on its capacity to do so are not abundant because until recently cultivation of Miscanthus was of most interest as a source of bioenergy, not remediation of contaminated sites. There are a limited number of greenhouse pot studies with selected compounds. These are discussed in reviews by Nsanganwimana et al. (2014) and Pidlisnyuk et al. (2014). By that date only the work of Técher et al. (2012a) reported a field study of significant organic [PAH] contamination. That same group showed that exudates of Miscanthus roots stimulated microbial degradation of some PAHs in microcosm studies (Técher et al., 2012b), while an earlier study showed enhancement of degradation of diesel fuel also (Técher et al., 2011).

One recent study (Wechtler et al., 2020) examined dissipation of PAHs from a technosol (mixture of dredged sediments and contaminated soil). Plants used were *M. ×giganteus*, white clover (*T. repens*), and a co-culture of the two. After a growing season of 263 days, there was a significant decrease of 16 priority PAHs, with ~30% decrease in the monocultures and co-culture compared to an unplanted technosol. This lowered the integrated average cancer risk from about 4.3 to 3.4. The co-culture also lowered the predicted ecotoxicity more than each monoculture, showing a greater decrease of anthracene and pyrene.

Miscanthus is known to have resistance to herbicides similar to that of maize (Anderson, 2011), but mechanisms are undefined, whether by metabolic tolerance, deactivation of the herbicide or exclusion from the plant. Anderson (2011) tested more than 20 herbicides, at several rates of application for control of broad-leaf weeds or other grasses (pre-emergence). Many herbicides were tested in greenhouse studies and some at small scale in a field.

Recalcitrance of PAHs and PCBs varies widely so that success with a few congeners or homologs is no guarantee of success with all forms. The earliest remediation study with Miscanthus that we can identify was by Wilke and Metz (1993). They analyzed a suite of six PCB congeners and six PAHs from a long-term contaminated site in Germany, which also had high levels of toxic metals including Cd and Zn relative to regulatory standards. The contamination source was sewage sludge irrigation and the soil total organic carbon was 6.7%. Growth of *Miscanthus sinensis (M. sinensis)* or *Polygonum sachalinense* was less effective in the undiluted soil containing 72 mg kg^{-1} of Cd and 1800 mg kg^{-1} of Zn, >1600 µg kg^{-1} of PCBs and 3062 µg kg^{-1} of PAHs than in a soil diluted with brown podzol. That soil had only 0.77% of TOC and a pH 4.1. It is not known which element or compound was the main contributor to growth inhibition. Depending on bioavailability, Zn at 1800 mg kg^{-1} at the pH < 5 (pH 4.8 in undiluted soil) could be strongly inhibitory. Over a range of dilutions giving 1×, 5×, 10×, 20× of the regulatory limit of 1.5 mg kg^{-1} for Cd, both PCBs and PAHs were accumulated only in roots, to a maximum of 975 µg kg^{-1} of PCBs and 2083 µg kg^{-1} of PAHs at approximately 1:1 dilution of the contaminated soil (to 20×1.5 mg kg^{-1} of Cd). Degradation products were not measured. Neither PCB nor PAH congeners were detected in stems and leaves in this pot study. No more recent studies with PCBs have been reported.

Little is known of the tolerance or degradation capacity of Miscanthus for POPs including chlorinated pesticides. One study was completed with the insecticide chlordecone (trade name Kepone) by Liber et al. (2018). This persistent material with a half-life of ~30 years may be present in soils at levels of >1 mg kg^{-1}. Using 14-C labeled material in a greenhouse study with dense planting of 49 plants in 0.81 m^2, accumulation above soil level (expressed as mg kg^{-1}) was observed only in roots of both *M. ×giganteus* (5 mg kg^{-1} in roots, 0.17 mg kg^{-1} in rhizomes, 0.15 mg kg^{-1} in shoots) and *M. sinensis* (15 mg kg^{-1} in roots, 0.5 mg kg^{-1} in rhizome, 0.3 mg kg^{-1} in shoots) during the second growing period of 2 and then 6 months. In a second experiment with *M. sinensis* over two growth periods of 10 months each, accumulation into aboveground tissues from 8 mg kg^{-1} in soil was less in the second period, to only about 2 mg kg^{-1}, presumably because of increased organic matter in the root zone. Calculated harvestable plant contamination under hypothetical field conditions (1 mg kg^{-1} soil) indicated about 1 g ha^{-1}year^{-1} would be removed by harvesting Miscanthus. With a contaminant load of perhaps 2 kg ha^{-1} (~1 mg kg^{-1}, ~20 cm depth). This is a trivial amount for recovery, though important for appropriate use of the harvested crop.

A study by Nurzhanova et al. (2017) presented the application of *M. sinensis* to phytoremediation of soil heavily polluted by DDT and its metabolites. The aged contaminated soil was sampled around a destroyed storehouse for chlorinated pesticides in Kazakhstan, and chemical analysis showed large pesticide concentrations in the soil. For example, DDT concentrations in the soil exceeded the maximum permissible concentration (MPC) limit value by 62 times. The results showed that *M. ×giganteus* could not develop in soil with a concentration of DDT higher than 45× MPC limit of DDT and its metabolites, while *M. sinensis* was able to develop in pesticide contaminated soil till 62× MPC. The observation showed (Table 3.1) that *M. ×giganteus* mainly accumulated the chlorinated pesticides in roots despite high pesticide concentration in the soil. With increasing concentration of DDT and its metabolites in the soil, the total uptake of 4,4'-DDE increased, uptake of 2,4'-DDD remained the same, and uptake of DDT decreased. In the case of *M. sinensis*, 4,4'-DDE and 4,4'-DDT were mainly accumulated in roots while 2,4'-DDD appeared more in aboveground biomass (Nurzhanova et al., 2017).

The phytoremediation potential of *M. sinensis* and the influence of two amendments, Tween 20 and AC added to the process, was studied with aged polluted soil from Kazakhstan (Mamirova et al., 2020). The soil contained heavy metals and 24 chlorinated pesticides (DDT and metabolites, HCH and isomers, endrin, keltan, aldrin, dieldrin, chlordane, and others) in concentrations that exceeded the MPC by 10–100 times (Table 3.2). Results showed that amendments changed *M. sinensis* physiological parameters. Specifically, Tween 20 increased the plant height by 16.6%, while AC increased it only by 3.6%, yet it was statistically significant. The opposite tendency was detected in the case of root dry mass. The addition of Tween 20 and AC decreased the mass by 64% and 49.7%, respectively, while the aboveground biomass increased by 6.6% in the presence of AC and decreased by 5% in the presence of Tween 20.

Ten out of twenty-four chlorinated pesticides present in the soil were found in Miscanthus biomass; however, only five translocated from roots to aboveground plant parts. When soil was amended with Tween 20, the number of pesticides taken up was six, while when the soil was amended by AC only four pesticides were taken up (Figure 3.1). Tween 20 increased the total uptake of pesticides except for 2,4'-DDD where uptake decreased by 38.7%. AC decreased the total uptake of pesticides by 46.6%–92.1% (Figure 3.1). The pesticides in plant tissues were distributed differently: γ-HCH and dieldrin mainly accumulated in the aboveground biomass while α-HCH, β-HCH, aldrin, 2,4'-DDD, and endrin accumulated in the roots. When plants were grown in contaminated soil without amendments, 4,4'-DDE and 4,4'-DDD were distributed almost equally in different plant parts, while 4,4'-DDT mainly accumulated in the leaves and stems. When contaminated soil was amended by Tween 20, the effect changed: 4,4'-DDE was mainly accumulated in aboveground biomass; 4,4'-DDD and 4,4'-DDT in the roots. The presence of AC affected the phytostabilization potential of *M. sinensis* in relation to 4,4'-DDE, 4,4'-DDD, 4,4'-DDT, i.e., they were mainly accumulated in the root system (Mamirova et al., 2020).

Calculation of the uptake index for ten chlorinated pesticides showed that in soil contaminated by pesticides without or with amendments, *M. sinensis* accumulated 4,4'-DDE more than 4,4'-DDT, followed by 4,4'-DDD which can be explained by high concentration of 4,4'-DDE in the studied soils and its bioavailability due to lower hydrophobicity level.

TABLE 3.1

The Concentration of DDT and Its Metabolites in the Research Soil, and Miscanthus Biomass after One Season of Phytoremediation[a]

Samples	Soil or Vegetation	Mass (kg)	Concentration (μg kg⁻¹) 4,4'-DDE	2,4'-DDD	4,4'-DDT	Uptake Index (μg) 4,4'-DDE	2,4'-DDD	4,4'-DDT
M. ×giganteus								
2 MPC	Initial soil	3	146 ± 22	3 ± 1	92 ± 21	438	9	276
	Aboveground	0.031	0.2 ± 0.1	18.3 ± 6.1	98.4 ± 27.1	62×10^{-4}	0.57	3.05
	Root	0.035	35.2 ± 6.5	352.3 ± 32.2	1810.4 ± 126.3	1.23	12.33	63.36
	Final soil	3	105 ± 18	2 ± 1	83 ± 12	315	6	249
M. ×giganteus Dry Biomass								
6 MPC	Aboveground	0.015	50.2 ± 14.2	15.3 ± 6.2	36.3 ± 12.5	0.75	0.23	0.55
	Root	0.026	158.1 ± 87.5	134.3 ± 25.1	139.8 ± 23.6	4.11	3.49	3.63
13 MPC	Aboveground	0.010	41.4 ± 21.2	17.3 ± 11.5	36.7 ± 15.7	0.41	0.17	0.37
	Root	0.022	246.3 ± 123.3	135.5 ± 27.1	191.1 ± 55.1	5.42	2.98	4.20
33 MPC	Aboveground	0.008	22.2 ± 9.2	36.1 ± 18.2	32.0 ± 16.0	0.18	0.29	0.26
	Root	0.020	267.4 ± 42.1	60.4 ± 19.1	212.6 ± 27.1	5.35	1.21	4.25
45 MPC	Aboveground	0.007	36.1 ± 27.2	112.3 ± 20.5	27.9 ± 9.9	0.25	0.79	0.20
	Root	0.021	453.3 ± 30.1	113.4 ± 15.2	52.1 ± 11.1	9.52	2.38	1.09
62 MPC	Aboveground	Died						
	Root							
M. sinensis Dry Biomass								
62 MPC	Initial soil	3	2750 ± 88	933 ± 48	2498 ± 45	8250	2799	7494
	Aboveground	0.014	151.2 ± 45.3	78.4 ± 29.3	12.4 ± 7.0	2.12	1.10	0.17
	Root	0.009	570.5 ± 53.3	45.1 ± 22.0	247.2 ± 76.4	5.13	0.41	2.22
	Final soil	3	1230 ± 49	888 ± 79	1991 ± 221	3690	2664	5973

Source: Modified from Nurzhanova et al. (2017).

[a]*Notice:* value 2 MPC refers to two times of MPC or $2 \times 0.1 \, mg \, kg^{-1} = 0.2 \, mg \, kg^{-1}$.

TABLE 3.2

Concentrations of Chlorinated Pesticides in the Aged Soil

Chlorinated Pesticides	MPC KZ,[a] ($\mu g\ kg^{-1}$)	MPC EU,[b] ($\mu g\ kg^{-1}$)	Concentration ($\mu g\ kg^{-1}$)
2,4'-DDD	100	10.0	14,072.0 ± 5,239.0
4,4'-DDD	100	10.0	11,434.0 ± 7,302.0
4,4'-DDE	100	10.0	777.9 ± 292.0
4,4'-DDT	100	10.0	10,023.0 ± 2,471
Aldrin	2.5	7.0	230.2 ± 59.1
Chlordane	100	4.3	48.1 ± 27.6
Chlorobenzilate	20	-	32,061.0 ± 12,669.0
Dibutyl chlorendate	-	-	2134.6 ± 477.6
Dieldrin	0.5	7.0	132.9 ± 51.1
Endosulfan α	100	0.003	5.5 ± 0.0
Endosulfan β	100	0.003	253.1 ± 163.1
Endosulfan sulfate	-	-	118.7 ± 76.5
Endrin	1	2.9	44,085.0 ± 17,335.0
Endrin aldehyde	-	2.9	1087.0 ± 198.0
HCB	500	50.0	4.7 ± 1.9
Heptachlor	50	0.7	214.7 ± 0.0
Heptachlorepoxide	50	0.052	3029.0 ± 1192.0
Hexabromobenzene	30	28.0	201.4 ± 129.9
Keltan (Dicofol)	100	-	34.4 ± 0.0
Methoxychlor	1600	900.0	435.6 ± 281.1
α-HCH	100	220.0	89.2 ± 0.0
β-HCH	100	92.0	25.5 ± 16.4
γ-HCH	100	0.01	488.0 ± 152.0
δ-HCH	100	-	67.4 ± 13.7

Source: Modified from Mamirova et al. (2020).

[a] Maximum Permissible Concentration (MPC) values for the Republic of Kazakhstan (MHRK & MEPRK, 2004).

[b] MPC values as for EU (Crommentuijn et al., 2000; Van de Plassche, 1994).

In conclusion, Miscanthus is likely to tolerate at least moderate levels of organic contaminants, unless they are specific plant growth regulators, or membrane disruptors. Whether it is able to metabolize particular organic compounds can only be determined confidently with plants of that genus. There are suggestions that different species or biovars and cultivars (CVs) may vary in capacity within a genus, in general, but there is little clear evidence with Miscanthus. The work of Mamirova et al. (2020) indicates there may be differences for some DDT metabolites. There will no doubt be other examples.

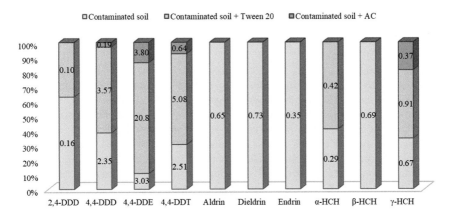

FIGURE 3.1
Total content of chlorinated pesticides in *M. sinensis*, units – μg (Modified from Mamirova et al. (2020).

References

Abdullah, S. R. S., Al-Baldawi, I. A., Almansoory, A. F., Purwanti, I. F., Al-Sbani, N. H., & Sharuddin, S. S. N. (2020). Plant-assisted remediation of hydrocarbons in water and soil: Application, mechanisms, challenges and opportunities. *Chemosphere, 247*, 125932. https://doi.org/10.1016/j.chemosphere.2020.125932.

Afzal, M., Khan, Q. M., & Sessitsch, A. (2014). Endophytic bacteria: Prospects and applications for the phytoremediation of organic pollutants. *Chemosphere, 117*(1), 232–242. https://doi.org/10.1016/j.chemosphere.2014.06.078.

Agbeve, S. K., Carboo, D., Duker-Eshun, G., Afful, S., & Ofosu, P. (2013). Burden of organochlorine pesticide residues in the root of *Cryptolepis sanguinolenta*, antimalarial plant used in traditional medicine in Ghana. *European Chemical Bulletin, 2*(11), 936–941. https://doi.org/10.17628/ECB.2013.2.936-941.

Agnello, A. C., Huguenot, D., Van Hullebusch, E. D., & Esposito, G. (2014). Enhanced phytoremediation: A review of low molecular weight organic acids and surfactants used as amendments. *Critical Reviews in Environmental Science and Technology, 44*(22), 2531–2576. https://doi.org/10.1080/10643389.2013.829764.

An, C. J., Huang, G. H., Wei, J., & Yu, H. (2011). Effect of short-chain organic acids on the enhanced desorption of phenanthrene by rhamnolipid biosurfactant in soil-water environment. *Water Research, 45*(17), 5501–5510. https://doi.org/10.1016/j.watres.2011.08.011.

Ancona, V., Barra Caracciolo, A., Grenni, P., Di Lenola, M., Campanale, C., Calabrese, A., Uricchio, V. F., Mascolo, G., & Massacci, A. (2017). Plant-assisted bioremediation of a historically PCB and heavy metal-contaminated area in Southern Italy. *New Biotechnology, 38*, 65–73. https://doi.org/10.1016/j.nbt.2016.09.006.

Anderson, E. K. (2011). Herbicide toxicity response and eradication studies in *Miscanthus × giganteus*, Thesis (M.S.) in Crop Science, University of Illinois. http://hdl.handle.net/2142/18498.

Arslan, M., Imran, A., Khan, Q. M., & Afzal, M. (2017). Plant–bacteria partnerships for the remediation of persistent organic pollutants. *Environmental Science and Pollution Research, 24*(5), 4322–4336. https://doi.org/10.1007/s11356-015-4935-3.

Asquer, C., Melis, E., Scano, E. A., & Carboni, G. (2019). Opportunities for green energy through emerging crops: Biogas valorization of *Cannabis sativa* L. residues. *Climate, 7*(12), 142. https://doi.org/10.3390/cli7120142.

Becerra-Castro, C., Prieto-Fernández, Á., Kidd, P. S., Weyens, N., Rodríguez-Garrido, B., Touceda-González, M., Acea, M. J., & Vangronsveld, J. (2013). Improving performance of *Cytisus striatus* on substrates contaminated with hexachlorocyclohexane (HCH) isomers using bacterial inoculants: Developing a phytoremediation strategy. *Plant and Soil, 362*(1–2), 247–260. https://doi.org/10.1007/s11104-012-1276-6.

Beesley, L., Moreno-Jiménez, E., Gomez-Eyles, J. L., Harris, E., Robinson, B., & Sizmur, T. (2011). A review of biochars' potential role in the remediation, revegetation and restoration of contaminated soils. *Environmental Pollution, 159*(12), 3269–3282. https://doi.org/10.1016/j.envpol.2011.07.023.

Bilalis, D., Karidogianni, S., Roussis, I., Kouneli, V., Kakabouki, I., & Folina, A. (2019). *Cannabis sativa* L.: A new promising crop for medical and industrial use. *Bulletin of University of Agricultural Sciences and Veterinary Medicine Cluj-Napoca. Horticulture, 76*(2), 145. https://doi.org/10.15835/buasvmcn-hort:2019.0020.

Cachada, A., Ferreira da Silva, E., Duarte, A. C., & Pereira, R. (2016). Risk assessment of urban soils contamination: The particular case of polycyclic aromatic hydrocarbons. *Science of the Total Environment, 551–552,* 271–284. https://doi.org/10.1016/j.scitotenv.2016.02.012.

Campanella, B. E., Bock, C., & Schröder, P. (2002). Phytoremediation to increase the degradation of PCBs and PCDD/Fs: Potential and limitations. *Environmental Science and Pollution Research, 9*(1), 73–85. https://doi.org/10.1007/bf02987318.

Campbell, S., Paquin, D., Awaya, J. D., & Li, Q. X. (2002). Remediation of benzo[a]pyrene and chrysene-contaminated soil with industrial hemp (*Cannabis sativa*). *International Journal of Phytoremediation, 4*(2), 157–168. https://doi.org/10.1080/15226510208500080.

Canadian Council Ministers Environments. (2008). Canada-wide standards for petroleum hydrocarbons (PHC) in soil. https://www.ccme.ca/en/resources/contaminated_site_management/phc_cws_in_soil.html

Chan-Quijano, J. G., Cach-Perez, M. J., & Rodriguez-Robles, U. (2020). Phytoremediation of soils contaminated by hydrocarbon. In: Shmaefsky, B. R. ed. *Phytoremediation, Concepts and Strategies in Plant Sciences*, Springer, New York, 83–101.

Chekol, T., Vough, L. R., & Chaney, R. L. (2004). Phytoremediation of polychlorinated biphenyl-contaminated soils: The rhizosphere effect. *Environment International, 30*(6), 799–804. https://doi.org/10.1016/j.envint.2004.01.008.

Chlebek, D., & Hupert-Kocurek, K. (2019). Endophytic bacteria in the phytodegradation of persistent organic pollutants. *Advancements of Microbiology, 58*(1), 70–79. https://doi.org/10.21307/PM-2019.58.1.070

Crommentuijn, T., Sijm, D., De Bruijn, J., Van Leeuwen, K., & Van de Plassche, E. (2000). Maximum permissible and negligible concentrations for some organic substances and pesticides. *Journal of Environmental Management, 58*(4), 297–312. https://doi.org/10.1006/jema.2000.0334.

Das, P. (2014). Chemically catalyzed phytoremediation of 2,4,6-trinitrotoluene (TNT) contaminated soil by vetiver grass (*Chrysopogon zizanioides* L.). *Theses, Dissertations and Culminating Projects*. Montclair State University. https://digitalcommons.montclair.edu/etd/57.

Dathan, J. (2020). The broken land: The environmental consequences of explosive weapon use. Action on Armed Violence, London, UK. www.aoav.org.uk.

Denyes, M. J., Langlois, V. S., Rutter, A., & Zeeb, B. A. (2012). The use of biochar to reduce soil PCB bioavailability to *Cucurbita pepo* and *Eisenia fetida*. *Science of the Total Environment, 437,* 76–82. https://doi.org/10.1016/j.scitotenv.2012.07.081.

Denyes, M. J., Parisien, M. A., Rutter, A., & Zeeb, B. A. (2014). Physical, chemical and biological characterization of six biochars produced for the remediation of contaminated sites. *Journal of Visualized Experiments : JoVE, 93,* e52183. https://doi.org/10.3791/52183.

Denyes, M. J., Rutter, A., & Zeeb, B. A. (2013). *In situ* application of activated carbon and biochar to PCB-contaminated soil and the effects of mixing regime. *Environmental Pollution, 182,* 201–208. https://doi.org/10.1016/j.envpol.2013.07.016.

Denyes, M. J., Rutter, A., & Zeeb, B. A. (2016). Bioavailability assessments following biochar and activated carbon amendment in DDT-contaminated soil. *Chemosphere, 144,* 1428–1434. https://doi.org/10.1016/j.chemosphere.2015.10.029.

Esteve-Núñez, A., Caballero, A., & Ramos, J. L. (2001). Biological degradation of 2,4,6-trinitrotoluene. *Microbiology and Molecular Biology Reviews, 65*(3), 335–352. https://doi.org/10.1128/mmbr.65.3.335-352.2001.

Fairchild, J. F., Ruessler, D. S., & Carlson, A. R. (1998). Comparative sensitivity of five species of macrophytes and six species of algae to atrazine, metribuzin, alachlor, and metolachlor. *Environmental Toxicology and Chemistry, 17*(9), 1830–1834. https://doi.org/10.1002/etc.5620170924.

Ficko, S. A., Rutter, A., & Zeeb, B. A. (2011). Phytoextraction and uptake patterns of weathered polychlorinated biphenyl-contaminated soils using three perennial weed species. *Journal of Environmental Quality, 40*(6), 1870–1877. https://doi.org/10.2134/jeq2011.0144.

Field, J. A., & Sierra-Alvarez, R. (2008). Microbial degradation of chlorinated dioxins. *Chemosphere, 71*(6), 1005–1018. https://doi.org/10.1016/j.chemosphere.2007.10.039.

Fiorenza, S., Oubre, C. L., & Ward, C. H. (2000). *Phytoremediation of Hydrocarbon Contaminated Soil*, Lewis Publishers, New York.

Gąbka, D., & Wolski, K. (2011). Use of turfgrasses in landfill leachate treatment. *Polish Journal of Environmental Studies, 20*(5), 1161.

Gandia-Herrero, F., Lorenz, A., Larson, T., Graham, I. A., Bowles, D. J., Rylott, E. L., & Bruce, N. C. (2008). Detoxification of the explosive 2,4,6-trinitrotoluene in *Arabidopsis*: Discovery of bifunctional O- and C-glucosyltransferases. *Plant Journal, 56*(6), 963–974. https://doi.org/10.1111/j.1365-313X.2008.03653.x.

Glick, B. R. (2010). Using soil bacteria to facilitate phytoremediation. *Biotechnology Advances, 28*(3), 367–374. https://doi.org/10.1016/j.biotechadv.2010.02.001.

Gonzalez, M., Miglioranza, K. S. B., Aizpún, J. E., Isla, F. I., & Peña, A. (2010). Assessing pesticide leaching and desorption in soils with different agricultural activities from Argentina (Pampa and Patagonia). *Chemosphere, 81*(3), 351–358. https://doi.org/10.1016/j.chemosphere.2010.07.021.

Hemapala, M. U. (2017). Robots for humanitarian demining. In: Canbolat, H. ed. *Robots Operating in Hazardous Environments*, Books on Demand, Norderstedt, Germany, 3–21. https://doi.org/10.5772/intechopen.70246.

Henner, P., Schiavon, M., Morel, J.-L., Lichtfouse, E., & Lichtfouse Polycyclic, E. (1997). Polycyclic aromatic hydrocarbon (PAH) occurrence and remediation methods. *Analysis, 25*(9). https://hal.archives-ouvertes.fr/hal-00193277.

Johnson, R. (2014). Hemp as an agricultural commodity. CRS Report No. RL32725. http://www.pahic.org/white-papers/.

Kalderis, D., Juhasz, A. L., Boopathy, R., & Comfort, S. (2011). Soils contaminated with explosives: Environmental fate and evaluation of state-of-the-art remediation processes (IUPAC technical eport). *Pure and Applied Chemistry, 83*(7), 1407–1484. https://doi.org/10.1351/PAC-REP-10-01-05.

Khalid, S., Shahid, M., Murtaza, B., Bibi, I., Natasha, Asif Naeem, M., & Niazi, N. K. (2020). A critical review of different factors governing the fate of pesticides in soil under biochar application. *Science of the Total Environment, 711*, 134645. https://doi.org/10.1016/j.scitotenv.2019.134645.

Kumar, S., Singh, R., Kumar, V., Rani, A., & Jain, R. (2017). *Cannabis sativa*: A plant suitable for phytoremediation and bioenergy production. In: Bauddh, K., Singh, B., Korstad, J. eds. *Phytoremediation Potential of Bioenergy Plants*, Springer, Singapore, 269–285. https://doi.org/10.1007/978-981-10-3084 0_10.

Liber, Y., Létondor, C., Pascal-Lorber, S., & Laurent, F. (2018). Growth parameters influencing uptake of chlordecone by Miscanthus species. *Science of the Total Environment, 624*, 831–837. https://doi.org/10.1016/j.scitotenv.2017.12.071.

Liu, J., & Schnoor, J. L. (2008). Uptake and translocation of lesser-chlorinated polychlorinated biphenyls (PCBs) in whole hybrid poplar plants after hydroponic exposure. *Chemosphere, 73*(10), 1608–1616. https://doi.org/10.1016/j.chemosphere.2008.08.009.

Loffredo, E., Picca, G., & Parlavecchia, M. (2020). Single and combined use of *Cannabis sativa* L. and carbon-rich materials for the removal of pesticides and endocrine-disrupting chemicals from water and soil. *Environmental Science and Pollution Research*, 1–16. https://doi.org/10.1007/s11356-020-10690-7.

Lunney, A. I., Zeeb, B. A., & Reimer, K. J. (2004). Uptake of weathered DDT in vascular plants: Potential for phytoremediation. *Environmental Science and Technology, 38*(22), 6147–6154. https://doi.org/10.1021/es030705b.

Mamirova, A., Pidlisnyuk, V., Amirbekov, A., Sevcu, A., & Nurzhanova, A. (2020). Phytoremediation potential of *Miscanthus sinensis* And. in organochlorine pesticides contaminated soil amended by Tween 20 and Activated carbon. *Environmental Science and Pollution Research*. https://doi.org/10.1007/s11356-020-11609-y.

McCutcheon, S. C., & Schnoor, J. L. (2003). *Phytoremediation: Transformation and Control of Contaminants*, Wiley, Hoboken, NJ.

Mench, M., Lepp, N., Bert, V., Schwitzguébel, J.-P., Gawronski, S. W., Schröder, P., & Vangronsveld, J. (2010). Successes and limitations of phytotechnologies at field scale: Outcomes, assessment and outlook from COST Action 859. *Journal of Soils and Sediments, 10*(6), 1039–1070. https://doi.org/10.1007/s11368-010-0190-x.

MHRK and MEPRK. (2004). Standards for maximum permissible concentrations of harmful substances, pests and other biological substances polluting the≈soil, approved by a joint order of the Ministry of Health of the Republic of Kazakhstan dated January 30, 2004 No. 99 and the Ministry of Environmental Protection of the Republic of Kazakhstan dated January 27, 2004, No. 21-P.

Narayanan, M., Davis, L. C., & Erickson, L. E. (1995). Fate of volatile chlorinated organic compounds in a laboratory chamber with alfalfa plants. *Environmental Science and Technology, 29*(9), 2437–2444. https://doi.org/10.1021/es00009a041.

Narayanan, M., Erickson, L. E., & Davis, L. C. (1999). Simple plant-based design strategies for volatile organic pollutants. *Environmental Progress, 18*(4), 231–242. https://doi.org/10.1002/ep.670180409.

Nartey, O. D., & Zhao, B. (2014). Biochar preparation, characterization, and adsorptive capacity and its effect on bioavailability of contaminants: An overview. *Advances in Materials Science and Engineering, 2014*, 715398. https://doi.org/10.1155/2014/715398.

Nichols, E. G., Cook, R. L., Landmeyer, J. E., Atkinson, B., Malone, D. R., Shaw, G., & Woods, L. (2014). Phytoremediation of a petroleum-hydrocarbon contaminated shallow aquifer in Elizabeth City, North Carolina, USA. *Remediation Journal, 24*(2), 29–46. https://doi.org/10.1002/rem.21382.

Nsanganwimana, F., Pourrut, B., Mench, M., & Douay, F. (2014). Suitability of Miscanthus species for managing inorganic and organic contaminated land and restoring ecosystem services. A review. *Journal of Environmental Management, 143*, 123–134. https://doi.org/10.1016/j.jenvman.2014.04.027.

Nurzhanova, A., Pidlisnyuk, V., Sailaukhanuly, Y., Kenessov, B., Trogl, J., Aligulova, R., Kalugin, S., Nurmagambetova, A., Abit, K., & Stefanovska, T. (2017). Phytoremediation of military soil contaminated by metals and organochlorine pesticides using miscanthus. *Communication in Agricultural and Applied Biological Sciences, 82*, 61–68.

Otani, T., Seike, N., & Sakata, Y. (2007). Differential uptake of dieldrin and endrin from soil by several plant families and *Cucurbita* genera. *Soil Science and Plant Nutrition, 53*(1), 86–94. https://doi.org/10.1111/j.1747-0765.2007.00102.x.

Paul, S., Rutter, A., & Zeeb, B. A. (2015). Phytoextraction of DDT-contaminated soil at Point Pelee National Park, Leamington, ON, using *Cucurbita pepo* cultivar Howden and native grass species. *Journal of Environmental Quality, 44*(4), 1201–1209. https://doi.org/10.2134/jeq2014.11.0465.

Pidlisnyuk, V., Stefanovska, T., Lewis, E. E., Erickson, L. E., & Davis, L. C. (2014). Miscanthus as a productive biofuel crop for phytoremediation. *Critical Reviews in Plant Sciences, 33*(1), 1–19. https://doi.org/10.1080/07352689.2014.847616.

Pillai, H. P. S., & Kottekottil, J. (2016). Nano-phytotechnological remediation of endosulfan using zero valent iron nanoparticles. *Journal of Environmental Protection, 7*(5), 734–744. https://doi.org/10.4236/jep.2016.75066.

Plhalova, L., Stepanova, S., Praskova, E., Chromcova, L., Zelnickova, L., Divisova, L., Skoric, M., Pistekova, V., Bedanova, I., & Svobodova, Z. (2012). The effects of subchronic exposure to metribuzin on *Danio rerio. The Scientific World Journal, 2012*, 728189. https://doi.org/10.1100/2012/728189.

Ramamurthy, A. S., & Memarian, R. (2012). Phytoremediation of mixed soil contaminants. *Water, Air, and Soil Pollution, 223*(2), 511–518. https://doi.org/10.1007/s11270-011-0878-6.

Rani, M., Shanker, U., & Jassal, V. (2017). Recent strategies for removal and degradation of persistent & toxic organochlorine pesticides using nanoparticles: A review. *Journal of Environmental Management, 190*, 208–222. https://doi.org/10.1016/j.jenvman.2016.12.068.

Ren, X., Zeng, G., Tang, L., Wang, J., Wan, J., Liu, Y., Yu, Y., Yi, H., Ye, S., & Deng, R. (2017) Sorption, transport and biodegradation – An insight into bioavailability of persistent organic pollutants in soil. *Science of the Total Environment, 610–611,* 1154–1163. https://doi.org/10.1016/j.scitotenv.2017.08.089.

Robledo, L., Carrasco, M., & Mery, D. (2009). A survey of land mine detection technology. *International Journal of Remote Sensing, 30*(9), 2399–2410. https://doi.org/10.1080/01431160802549435.

Saibu, S., Adebusoye. S. A., & Oyetibo, G. O. (2020) Aerobic bacterial transformation and biodegradation of dioxins: a review. *Bioresources and Bioprocessing, 7,* 7. https://doi.org/10.1186/s40643-020-0294-0.

Santharam, S., Ibbini, J., Davis, L. C., & Erickson, L. E. (2011). Field study of biostimulation and bioaugmentation for remediation of tetrachloroethene in groundwater. *Remediation Journal, 21*(2), 51–68. https://doi.org/10.1002/rem.20281.

Santoyo, G., Moreno-Hagelsieb, G., del Carmen Orozco-Mosqueda, M., & Glick, B. R. (2016). Plant growth-promoting bacterial endophytes. *Microbiological Research, 183,* 92–99. https://doi.org/10.1016/j.micres.2015.11.008.

Squillace, P. J., Pankow, J. F., Korte, N. E., & Zogorski, J. S. (1997). Review of the environmental behavior and fate of methyl *tert*-butyl ether. *Environmental Toxicology and Chemistry, 16*(9), 1836–1844. https://doi.org/10.1002/etc.5620160911.

Tang, K. H. D. (2019). Phytoremediation of soil contaminated with petroleum hydrocarbons: A review of recent literature. *Global Journal of Civil and Environmental Engineering, 1,* 33–42. https://doi.org/10.36811/gjcee.2019.110006.

Tarla, D. N., Erickson, L. E., Hettiarachchi, G. M., Amadi, S. I., Galkaduwa, M., Davis, L. C., Nurzhanova, A., & Pidlisnyuk, V. (2020). Phytoremediation and bioremediation of pesticide-contaminated soil. *Applied Sciences, 10*(4), 1217. https://doi.org/10.3390/app10041217.

Técher, D., D'Innocenzo, M., Laval-Gilly, P., Henry, S., Bennasroune, A., Martinez-Chois, C., & Falla, J. (2012a). Assessment of *Miscanthus × giganteus* secondary root metabolites for the biostimulation of PAH-utilizing soil bacteria. *Applied Soil Ecology, 62,* 142–146. https://doi.org/10.1016/j.apsoil.2012.06.009.

Técher, D., Laval-Gilly, P., Henry, S., Bennasroune, A., Formanek, P., Martinez-Chois, C., D'Innocenzo, M., Muanda, F., Dicko, A., Rejšek, K., & Falla, J. (2011). Contribution of *Miscanthus × giganteus* root exudates to the biostimulation of PAH degradation: An *in vitro* study. *Science of the Total Environment, 409*(20), 4489–4495. https://doi.org/10.1016/j.scitotenv.2011.06.049.

Técher, D., Martinez-Chois, C., Laval-Gilly, P., Henry, S., Bennasroune, A., D'Innocenzo, M., & Falla, J. (2012b). Assessment of *Miscanthus × giganteus* for rhizoremediation of long term PAH contaminated soils. *Applied Soil Ecology, 62,* 42–49. https://doi.org/10.1016/j.apsoil.2012.07.009.

Terzaghi, E., Alberti, E., Raspa, G., Zanardini, E., Morosini, C., Anelli, S., Armiraglio, S., & Di Guardo, A. (2021). A new dataset of PCB half-lives in soil: Effect of plant species and organic carbon addition on biodegradation rates in a weathered contaminated soil. *Science of the Total Environment, 750,* 141411. https://doi.org/10.1016/j.scitotenv.2020.141411.

USEPA. (2015). Revegetating landfills and waste containment areas fact sheet. https://www.epa.gov/sites/production/files/2015-08/documents/revegetating_fact_sheet.pdf.

USEPA. (n.d.). Polychlorinated biphenyls (PCBs). https://www.epa.gov/pcbs/learn-about-polychlorinated-biphenyls-pcbs.

Van de Plassche, E. J. (1994). Towards integrated environmental quality objectives for several compounds with a potential for secondary poisoning. RIVM Rapport 679101012.

Vetiver Network International. (2017). Vetiver system for landfill leachate treatment. http://www.vetiver.org/TVN_VS_GAL_PUB/VS_Landfill_leachate_o.pdf.

Via, S. M. (2020). Phytoremediation of explosives. In: Shmaefsky, B. R. ed. *Phytoremediation: Concepts and Strategies in Plant Sciences*, Springer, New York, 261–284.

Wechtler, L., Henry, S., Falla, J., Walderdorff, L., Bonnefoy, A., & Laval-Gilly, P. (2020). Polycyclic aromatic hydrocarbons (PAHs) dissipation from a contaminated technosol composed of dredged sediments with *Miscanthus × giganteus* and *Trifolium repens* L. in mono- and co-culture. *Journal of Soils and Sediments, 20,* 2893–2902. https://doi.org/https://doi.org/10.1007/s11368-020-02648-6.

White, J. C., & Kottler, B. D. (2002). Citrate-mediated increase in the uptake of weathered 2,2-bis(*p*-chlorophenyl)1,1-dichloroethylene residues by plants. *Environmental Toxicology and Chemistry, 21*(3), 550–556. https://doi.org/10.1002/etc.5620210312.

Wilke, B.-M., & Metz, R. (1993). Soil/Plant Transfer of Pollutants by Cultivation of Energy Plants on Waste Water Irrigated Soils. In: Arendt, F. et al., eds. *Contaminated Soil '93*, Kluwer Publishing Co, Dordrecht, Netherlands, 581–582. https://doi.org/10.1007/978-94-011-2018-0_112.

4

Phytomining Applied for Postmining Sites

Hermann Heilmeier

Abstract

The economics associated with establishing and growing vegetation at contaminated sites can be improved if a metal that has commercial value can be extracted from the soil by plants. Phytomining is the process of extracting a product such as nickel from soil using hyperaccumulator plants that are able to grow in the contaminated soil and accumulate a metal product of value. After harvesting the plant biomass, drying it, and burning it for energy recovery, the ash can be processed to extract the metal of interest. Nickel and gold are good examples of products that have commercial value when phytomining is implemented at a contaminated site. Solar energy is used in phytomining, and soil quality is improved in many cases by increasing soil organic carbon and improving biological health and diversity in the soil. This chapter includes a review of phytomining and an analysis of its applications at contaminated sites containing metals that have commercial value.

CONTENTS

4.1 Introduction

Phytomining uses the capacity of plants and their associated microorganisms to extract and accumulate trace elements at high concentrations in their (above-)ground biomass (phytoextraction). However, in contrast to applications of phytoextraction for removing toxic trace elements (e.g., heavy metals, arsenic, selenium) from the environment as one option of phytoremediation (Pilon-Smits, 2005), the aim of phytomining processes is to extract commercially valuable elements from sources where the concentration of these elements is too low for economic activities applying conventional mining technologies, such as subeconomic ore bodies, mineral wastes (mine tailings), metal-bearing, or polluted soils, e.g., in (post-)mining areas (Heilmeier & Wiche, 2020; Naila et al., 2019; Sheoran et al., 2009). Most prominent examples in phytomining applied "hyperaccumulator" plants, particularly for nickel (e.g., Anderson et al., 1998; Kidd et al., 2018; Li et al., 2003) or gold (e.g., Wilson-Corral et al., 2011). The term "hyperaccumulator" had been proposed by Brooks et al. (1977) for plant taxa that accumulate above 1000 mg kg^{-1} of nickel (Ni) in their aboveground dry biomass. In the meantime, the term "hyperaccumulator" has been applied to a number of different elements, with respective adjustments of the concentration level (Jaffré et al., 2018; Rascio & Navari-Izzo, 2011; Van der Ent et al., 2013). Hyperaccumulator plant species were first suggested by Chaney (1983) for the purpose of phytomining. Nicks & Chambers (1995, 1998) were the first to perform field trials on phytomining of Ni, using the naturally occurring Ni-hyperaccumulator *Streptanthus polygaloides*. Brooks et al. (1998) argued that phytomining of Ni should be generally feasible due to a number of plants which accumulate Ni to high shoot concentrations (>10,000 mg kg^{-1}) and produce high biomass (> 10 t ha^{-1}). Soon, other elements such as thallium, copper, cobalt, and particularly gold have been tested for their phytomining potential (Anderson et al., 1998; Sheoran et al., 2013; Wilson-Corral et al., 2011, 2012).

In this chapter advantages and limitations of phytomining will be discussed, followed by a description of field experiments, particularly for phytomining of gold and Ni applying the so-called "hyperaccumulator plants" and recommendation for agronomic practices as derived from these experiments. Problems of economic viability and environmental implications will be addressed, based on the outcomes of various modelling studies. Examples on options of commercial application include phytomining for Ni from ultramafic soils, production of nanoparticles as catalysts, and extraction of valuable elements such as rare earth elements (REEs) from secondary sources for raw materials such as mine tailings.

4.2 Advantages and Limitations of Phytomining

In comparison to conventional mining technologies phytomining has several advantages. The two most important advantages are the fact that phytomining offers the option to exploit ores that are not economic for conventional mining approaches (Anderson et al., 1999), and the low cost of operation (Robinson et al., 2003). Further advantages are the use of cheap solar energy for creating a "bio-ore", low energy inputs for melting, less SO_x emissions due to bio-ore being sulfur-free, the improvement of site quality and other synergy effects with related industries as a consequence of biomass growth, e.g., generation of bioenergy (combustion, fermentation, generation of heat and biofuels), and the public acceptance as a "green" technology (Ali et al., 2013; Harris et al., 2009; Koptsik, 2014; Robinson et al., 2003; Saxena et al., 2020). With respect to the environment the sequestration of atmospheric carbon dioxide (CO_2) in plant biomass and due to enrichment of soils with organic matter and other ecosystem effects such as increased soil biodiversity, improved agronomic crop productivity, land restoration and pollution control are of relevance, which have to be considered in complete economic analyses (Kidd et al., 2018; Saxena et al., 2020). Thus, already around the turn of the century, Li et al. (2003) pointed to the option of sales of carbon credits. Robinson et al. (2009) emphasized the positive effects of phytomining in the remediation of sites polluted and degraded, e.g., due to mining and metallurgical processes, tailings, dumps, etc. The reduction of erosion (wind, surface runoff) and leaching of toxic substances to groundwater will improve water quality (Saxena et al., 2020). In summary, phytoremediation is "safe, aesthetic, nonaggressive, nondestructive" (Koptsik, 2014).

However, there are also severe limitations and restrictions negatively affecting the general applicability and success of phytomining. The most important limitation is due to the plants' shallow rooting system which allows for minerals close to the surface only to be extracted via this "green" technology (Hunt et al., 2014; Robinson et al., 2003). Furthermore, adverse site conditions, e.g., poor physical and chemical properties of soils on contaminated sites, usually limit plant growth severely; most stressful soil factors are the low content of nutrients and soil organic matter, and high soil acidity leading to solution of heavy metals (Hunt et al., 2014; Koptsik, 2014). Koptsik (2014) and Saxena et al. (2020) point out that for climatic and seasonal reasons, phytomining is more suitable in tropical and subtropical climates. Robinson et al. (2003, 2009) emphasize that phytomining is a long-term process with a much larger demand for area per unit of valuable elements compared to conventional mining and therefore may cause huge environmental disturbances in case, e.g., of clearing of natural vegetation for phytomining. Even if the total concentration of target elements in the soil may be high, their availability to the plants is often too low for effective uptake (Heilmeier & Wiche, 2020; Robinson et al., 2009; Sheoran et al., 2009). Hyperaccumulators with high uptake rates often show a low biomass production (Hunt et al., 2014; Robinson et al., 2003). Thus, two options for increasing element uptake of nonaccumulator plants have been proposed: (i) genetically modified plants (e.g., Hunt et al., 2014; Koptsik, 2014; Saxena et al., 2020); (ii) "induced hyperaccumulation" via applying chelating soil amendments (e.g., Hunt et al., 2014; Koptsik, 2014; Robinson et al., 2003; Wang et al., 2020; Wilson-Corral et al., 2012).

4.3 Field Experiments on Phytomining

Whereas part of the initial experiments on phytomining had been conducted under totally artificial conditions in the laboratory, applying, e.g., solubilizing agents such as ammonium thiocyanate for phytoextraction of gold (e.g., Anderson et al., 1998), systematic field trials, following the pioneering work of Nicks & Chambers (1995, 1998) and Robinson et al. (1997a, b) on Ni, have been performed by Anderson et al. (2005) for gold (Au), testing the two plant species *Brassica juncea* (Indian mustard) and *Zea mays* (corn), commonly used for phytoremediation and as energy crops. Applying an empirical model on the relationship between Au concentration in the soil and in plants obtained from laboratory and greenhouse studies, they suggested a minimum Au concentration in the substrate of 2 mg kg^{-1} to achieve a gold concentration in crops of 100 mg kg^{-1}, which, given a biomass harvest of 10 t ha^{-1}, would yield an economically viable gold recovery of 1 kg in plants per hectare. The authors proposed reclamation of spent heap-leach piles or retreatment of waste dumps, e.g., from artisanal gold mining as possible applications. Later on, Wilson-Corral et al. (2011), by application of chemical amendments, achieved average Au concentrations in leaves and stems of *Helianthus annuus* (sunflower) up to 19 and 21 mg kg^{-1}, still being well below the economic threshold proposed by Anderson et al. (2005). A field experiment to assess phytomining feasibility for artisanal gold mining tailings with tobacco plants, applying NaCN as chelating agent, however yielded Au concentrations of 1.2 mg kg^{-1} dry leaf biomass only (Krisnayanti et al., 2016).

The most promising element for phytomining at present is Ni (Kidd et al., 2018; Nkrumah et al., 2016). In a large number of experiments some ten plant species, with a focus on *Alyssum murale* (syn. *Odontarrhena chalcidica*) (e.g., Matko Stamenković et al., 2017; Rosenkranz et al., 2019), have been tested for their phytomining potential on ultramafic or Ni contaminated soils in different parts of the world. A major outcome from these studies is the positive effect of fertilization with nitrogen (N), phosphorus (P), and potassium (K) and of addition of organic matter on biomass of Ni hyperaccumulating plants, and the increase of Ni uptake and accumulation in shoots by adjustment of soil pH (optimum pH 5–7), addition of sulfur (S), or inoculation with (rhizo-)bacteria and mycorrhiza (Kidd et al., 2018; Nkrumah et al., 2016; Rosenkranz et al., 2019).

4.4 Agronomic Practices

As shown above for gold, the application of synthetic solubilizing agents greatly enhances Au uptake and accumulation in the aboveground biomass (Anderson et al., 1998; González-Valdez et al., 2018; Wilson-Corral et al., 2011). However, apart from the high costs of chelators, chemically induced hyperaccumulation, as already pointed out by Robinson et al. (2003), bears a number of serious environmental risks, such as persistence time of the chelators in the environment, leaching of mobilized toxic trace elements to ground water, or toxicity to plants and their associated microbes being used for phytomining (Hunt et al., 2014; Robinson et al., 2009; Saxena et al., 2020; Wang et al., 2020; Wilson-Corral et al., 2011). For that reason, more "natural" approaches such as soil management via conventional agricultural technologies have been initiated already among the first field trials, e.g., management of pH and fertilization (both inorganic, e.g., NPK, and organic, e.g., composted sewage sludge) for enhancing plant growth and phytoaccumulation of Ni (Chaney et al., 2007; Kidd et al., 2018; Li et al., 2003). Later on, co-cropping has been adopted for enhancing phytoextraction (Tang et al., 2012). Due to the low availability of target elements often limiting phytomining success (Heilmeier & Wiche, 2020; Sheoran et al., 2009), the stimulation of biological activity in the soil, particularly the rhizosphere, seems to be a most promising approach (Robinson et al., 2009). Apart from enhancing biomass production by Plant Growth Promoting Rhizobacteria, the exudation of metabolites such as organic acids and metal-chelating siderophores by soil microorganisms may change speciation of trace elements and thus greatly increase their solubility and bioavailability in the rhizosphere (Kidd et al., 2018; Koptsik, 2014; Saxena et al., 2020; Wiche et al., 2017).

Plant-targeted approaches include both traditional breeding for combining agronomic traits relevant for successful phytoaccumulation (Hunt et al., 2014; Li et al., 2003; Nkrumah et al., 2016; Robinson et al., 2009) or genetic approaches (Koptsik, 2014; Li et al., 2003), particularly for metal transporters (Hunt et al., 2014; Robinson et al., 2009). The most important criteria when selecting plant species for phytomining are as follows (Hunt et al., 2014; Koptsik, 2014; Li et al., 2003; Nkrumah et al., 2016; Saxena et al., 2020):

- easy cultivation as an agricultural crop, e.g., high rates of germination and establishment, easy propagation via seeds or cuttings
- adaptation to site climatic and edaphic conditions
- resistance to diseases and pests
- rapid growth
- high biomass yield
- extensive root system (deep, highly branched)
- high tolerance of elevated concentrations of (toxic) trace elements and extreme soil properties (pH, salinity)
- high specificity for target element(s)
- high uptake and translocation of target element(s) from roots to shoot
- high accumulation potential for target element(s) in aboveground plant parts
- potential for use as energy crop (burning, fermentation)

No single plant species will show all these traits; however, collection of germplasm from a diversity of accumulator plant species, testing potentials for agronomic yield and element accumulation on a variety of substrates with various soil management, and either conventional breeding or genetic manipulation should contribute to generate optimum "phytominers".

In conclusion, from an agroecosystem point of view, a combination of the following practices is the most promising approach (Kidd et al., 2018; Li et al., 2003; Nkrumah et al., 2016; Robinson et al., 2009; Tang et al., 2012; Wang et al., 2020):

- soil management
  - soil amendments (regulation of pH, inorganic, and organic fertilizers)
  - ploughing → decrease of soil heterogeneity (patchiness of concentration of target elements); translocation of target elements from deeper soil horizons close to soil surface
- crop management
  - planting: seeding depth, seed bed preparation, germination requirements, direct seeding, seed pelletization, transplanting
  - planting density (dependent on soil physical properties, organic matter and nutrient levels, plant species)
  - plant life form (annual, perennial)
  - co-cropping/intercropping → affects conditions in the rhizosphere → increased bioavailability of target element(s)
  - weed control
  - harvest methods: machinery for aboveground biomass, time of harvesting (effect on element and bioenergy yield); double harvesting
  - plant breeding

Widening the scope from the earlier focus on (hyper-)accumulators to the whole soil–plant agrosystem has been mirrored in the term "agromining" (Nkrumah et al., 2018). "Metal crops", i.e., plant species accumulating heavy metals, precious elements, or REEs, with a high biomass yield, have now become the focus of research (Li et al., 2020).

4.5 Economic Viability and Environmental Considerations

The agro-ecological expenditures related to plant cultivation as discussed above are only part of the costs of a complete phytomining approach, which also include processing of the plant biomass for recovery of the economically valuable elements. First crude economic analyses of phytomining, based on chelate-induced phytoextraction for increasing the solubility of, e.g., gold (Au), considered the costs for synthetic ligands added to the soil only (Anderson et al., 1998). Later on, more advanced economic models, including specified costs for site preparation, seeding, plant cultivation, application of fertilizers and chelating agents, and harvesting, yielded that expenditures for gold recovery by solvent extraction are the most dominant costs, which, however, can be dramatically reduced by burning biomass, thus reducing the volume of material for solvent extraction (Anderson et al., 2003). Burning of dry plants (e.g., *Odontarrhena muralis* with 1% Ni in the plant mass) yielded a concentration factor for Ni of ca. 12 (Kidd et al., 2018). Thus, depending on the initial plant concentration of Ni (1%–2% in several hyperaccumulators of the *Brassicaceae* family), Ni concentration in the ash may reach values of 10%–20% which is much more than in Ni ore mined from laterites (Simonnot et al., 2018). Robinson et al. (2003) developed a detailed mathematical model for assessing economic viability of phytomining as a function of multiple variables (costs for planting and producing biomass, value of biomass, and bio-ore) including interest rates and compared phytoextraction with alternative technologies. Applying their model to phytomining of Ni and Au in Australia, Harris et al. (2009) concluded that the most decisive factors for profitability are metal prices and the content of extractable metal(s). Another application of the same model by Wilson-Corral et al. (2012) concluded that phytomining for Au should be economically lucrative for metalliferous or abandoned mine sites. For less precious metals than gold, such as Ni, Van der Ent et al. (2015) emphasized the added market value of Ni catalysts for organic chemistry or pure Ni salt crystals.

An important issue neglected in early cost-benefit analyses is related to generation of bioenergy, e.g., via fermentation or combustion, and to increase of soil organic matter and sequestration of atmospheric CO_2 particularly on infertile marginal soils, the sale of carbon dioxide credits as already pointed out by Li et al. (2003). Carbon credits are part of a more holistic environmental evaluation of phytomining as proposed by Van der Ent et al. (2015). Kidd et al. (2018) advocated Life Cycle Assessment (LCA) as the most recognized method with many applications in agriculture and phytoremediation (see literature cited in Kidd et al., 2018). One of the first applications of LCA in the framework of phytomining was performed by Rodrigues et al. (2016), who considered, among others, toxicity to humans and environmental pollution caused by soil erosion (e.g., eutrophication) as a consequence of nonconservative agriculture. Erosion control, e.g., by contour farming or winter cover crops, can reduce not only loss of valuable metals but also impairment of human and ecosystem health. In addition, the use of biomass of accumulator plants not only reduces costs of phytomining but avoids atmospheric CO_2 increase by substituting fossil fuels. Already 1 year earlier, Echevarria et al. (2015) had emphasized a number of ecosystem services such as amelioration of soil quality, production of biofuel, conservation and restoration of biodiversity (protection of rare and endangered species, support of pollinating insects), and enhancement of carbon storage in soils. Apart from these positive ecological and environmental effects, Kidd et al. (2018) critically discuss also possible negative effects of phytomining on biodiversity (introduction of exotic species, behavioral consequences for pollinators by metal contents of flowers), soil CO_2 emissions as dependent on agricultural management, depletion of natural resources (e.g., water in case of irrigation), and the natural environment (land use and land use change), and emphasize the need for more research on descriptors and indicators, assessment methods, and cause-and-effect chain models on ecosystem services (Bouma & van Beukering, 2015; Kumar, 2010).

4.6 Options for Commercial Application of Phytomining

Although phytomining has a number of technological, economic, and environmental advantages, among others due to limitations as discussed in Section 4.2 (limited soil volume explored by plant rooting systems, adverse site conditions, low bioavailability of target elements), there is limited experience with field experiments (see Section 4.3) and technological applications. The most well-known and popular application of phytoremediation is the use of *Alyssum murale* for extracting Ni from ultramafic soils at the shores of Lake Ohrid in Albania, e.g., by the French company Econick (2018), based on long-term research on improving phytomining efficiency, e.g., by fertilization, weed control, and planting techniques (Bani et al., 2015). Ni recovered by hyperaccumulator plants can be processed, e.g., via hydrometallurgical technologies which have already been developed up to pilot scale for the production of Ni salts (ammonium and nickel sulfate hexahydrate, nickel sulfate, nickel acetate, etc. (Simonnot et al., 2016).

Already 10 years earlier, Haverkamp et al. (2007) suggested to synthesize metal nanoparticles by plants for catalytic purposes. Harumain et al. (2017) tested the suitability of plant species from various growth forms (mustard, miscanthus, willow) to extract palladium (Pd) from mine-sourced tailings. Although the accumulation of Pd was still below the target for commercially available 3% Pd-on-carbon catalysts, authors emphasize the strong potential for supplementary Pd supply by phytomining and positive environmental effects due to re-vegetation of tailings and other areas contaminated by mining activities and the restoration of their ecosystem functions.

The potential of both woody (e.g., *Populus tremula*) and nonwoody (e.g., *Phragmites australis* and *Phalaris arundinacea*) plant species for extracting valuable elements such as germanium and REEs from a dump field has also been demonstrated by Midula et al. (2017). An alternative source for conservation of primary resources by phytoextraction was suggested by Rosenkranz et al. (2017): waste incineration bottom ash. "Exotic" elements investigated so far were studied by Novo et al. (2015) for phytomining potential of Rhenium with Indian mustard (expected profit ca. 4000 US-$ ha^{-1}) and Shi et al. (2020) for platinum group metals to obtain nanoparticles, both under controlled conditions.

One of the earliest extended field experiments on phytoremediation of a former Uranium mining site (plot size 2 m×2 m, 1 m deep) was performed by Willscher et al. (2013) at Ronneburg (Thuringia, Germany). Soil improvement (amendment with calcareous top soil, mycorrhiza+bacteria) reduced the concentration of contaminants in seepage water, and rates and loads of seepage water. Neither ethanol fermentation nor biogas production was inhibited by uranium or heavy metals accumulated in low concentrations in the plant biomass; thus, the plant material from phytoremediation could be used for winning of bioenergy.

Another large-scale experiment for remediation of polluted soils via phytotechnologies has been carried out in Southwestern Europe (PhytoSUDOE, 2019). Nonfood crops, supported by soil amendments such as compost and bioaugmentation (inoculation with beneficial microorganisms), have been cultivated for rehabilitation of contaminated sites and production of useable biomass in Portugal, Spain, and France.

As shown by Harumain et al. (2017) and Midula et al. (2017), mine tailings are a huge reservoir of secondary resources, not only of elements considered traditionally in phytomining such as gold and nickel but also of less common elements such as palladium (Pd), germanium (Ge), and REEs. The phytomining potential for Ge and REEs has been investigated for a number of herbaceous plant species (both forbs and grasses) at experimental field sites close to Freiberg (Saxony, Germany), with the option to generate bioenergy either via fermentation or combustion. Whereas grasses such as *Ph. arundinacea*, *Avena sativa*, or *Zea mays* proved to be good accumulators for Ge due to their high capacity for uptake of the chemically similar Silicium, forbs like *Fagopyrum esculentum* or *Brassica napus* turned out to accumulate high amounts of REEs (Wiche & Heilmeier, 2016). Intercropping of *A. sativa* (oat) with *Lupinus albus* (white lupine), a leguminous plant with a high capacity for exudation of organic acids (e.g., citric acid) which can increase bioavailability of elements in the soil, increased accumulation of REEs in oat (Wiche et al., 2016). According to an economic analysis by Rentsch et al. (2016), both the accumulation of target elements in the plants and the high costs of extraction of target elements from fly ash after combustion of fermentation residues from biogas production are key factors for economic feasibility.

4.7 Conclusions and Perspectives

Although there are still major challenges to be passed for an economically viable application of phytomining on a broad scale, such as selection and improvement of promising target plants and the low bioavailability of target elements in the rooting environment, there are already some successful approaches, particularly for the element Ni, applying naturally occurring hyperaccumulator plants on ultramafic soils. Bio-ores produced via phytomining have a number of advantages compared to mineral ores, such as low energy demand for melting and less emissions of sulfur dioxide. Furthermore, plants have the capacity to synthesize nanoparticles with a high catalytic and absorptive activity. Nevertheless, optimizing processing of bio-ores for recovery of target elements and an improved understanding of plant–microbe–element interactions and stimulation of rhizosphere processes (e.g., via co-cropping) to increase bioavailability and thus accumulation of target elements in plants will be key parameters for economic viability of phytomining at the individual business company level. Economic return can be increased by utilizing bioenergy (fermentation, combustion) from accumulator plants; however, this requires more breeding efforts considering the low biomass yield of most (hyper)accumulator plants. Returning fermentation residues from biogas production as fertilizers to field sites will not only increase biomass yield of accumulator plants via fertilizing effects but also close nutrient loops (circular economy). Furthermore, application of organic matter from fermentation residues on marginal sites will improve soil conditions, reduce soil erosion, and thus contribute to soil and (ground-)water protection. In addition, sequestration of atmospheric CO_2 due to long-term soil improvement allows sales of carbon dioxide credits, which will not only increase financial returns on a microeconomic level but should also be included as positive effects of phytomining at the macroeconomic level as part of a more holistic economic and ecological evaluation of mining.

References

Ali, H., Khan, E., & Sajad, M. A. (2013). Phytoremediation of heavy metals—Concepts and applications. *Chemosphere, 91*(7), 869–881. https://doi.org/10.1016/j.chemosphere.2013.01.075.

Anderson, C. W. N., Brooks, R. R., Chiarucci, A., Lacoste, C. J., Leblanc, M., Robinson, B. H., Simcock, R., & Stewart, R. B. (1999). Phytomining for nickel, thallium and gold. *Journal of Geochemical Exploration, 67*(1–3), 407–415. https://doi.org/10.1016/S0375-6742(99)00055-2.

Anderson, C. W., Brooks, R. R., Stewart, R. B., & Simcock, R. (1998). Harvesting a crop of gold in plants. *Nature, 395*(6702), 553–554. https://doi.org/10.1038/26875.

Anderson, C., Moreno, F., & Meech, J. (2005). A field demonstration of gold phytoextraction technology. *Minerals Engineering, 18*(4), 385–392. https://doi.org/10.1016/j.mineng.2004.07.002.

Anderson, C. W. N., Stewart, R. B., Moreno, F. N., Wreesmann, C. T. J., Gardea-Torresdey, J. L., Robinson, B. H., & Meech, J. A. (2003). Gold phytomining. Novel developments in a plant-based mining system. *Proceedings of the Gold 2003 Conference: New Industrial Applications of Gold.*

Bani, A., Echevarria, G., Sulçe, S., & Morel, J. L. (2015). Improving the agronomy of *Alyssum murale* for extensive phytomining: A five-year field study. *International Journal of Phytoremediation, 17*(2), 117–127. https://doi.org/10.1080/15226514.2013.862204.

Bouma, J. A., & van Beukering, P. J. H. (2015). *Ecosystem Services: From Concept to Practice.* Cambridge University Press, Cambridge, UK. ISBN: 9781107062887.

Brooks, R. R., Lee, J., Reeves, R. D., & Jaffre, T. (1977). Detection of nickeliferous rocks by analysis of herbarium specimens of indicator plants. *Journal of Geochemical Exploration, 7*(C), 49–57. https://doi.org/10.1016/0375-6742(77)90074-7.

Brooks, R. R., Chambers, M. F., Nicks, L. J., & Robinson, B. H. (1998). Phytomining. *Trends in Plant Science, 3*(9), 359–362. https://doi.org/10.1016/S1360-1385(98)01283-7.

Chaney, R. L. (1983). Plant uptake of inorganic waste. In: Parr, J. F., Marsh, P. B., Kla, J. M. eds. *Land Treatment of Hazardous Wastes,* 50–76. Noyes Data Corporation, Park Ridge, NJ. ISBN: 10: 081550926X.

Chaney, R. L., Angle, J. S., Broadhurst, C. L., Peters, C. A., Tappero, R. V., & Sparks, D. L. (2007). Improved understanding of hyperaccumulation yields commercial phytoextraction and phytomining technologies. *Journal of Environmental Quality, 36*(5), 1429–1443. https://doi.org/10.2134/jeq2006.0514.

Echevarria, G., Baker, A., Morel, J.-L., Van Der Ent, A., Houzelot, V., Laubie, B., Pons, M.-N., Simonnot, M.-O., Zhang, X., Kidd, P., Benizri, E., Louis, J., Pons, M.-N., & Bani, A. (2015). Agromining for nickel: A complete chain that optimizes ecosystem services rendered by ultramafic landscapes. *13th SGA Meeting "Mineral Resources in a Sustainable World, Society for Geology Applied to Mineral Deposits (SGA). DEU.* ASGA Association Scientifique Géologie & Applications. https://hal.archives-ouvertes.fr/hal-01263575.

Econick. (2018). https://www.econick.fr/fr (last accessed October 29, 2020).

González-Valdez, E., Alarcón, A., Ferrera-Cerrato, R., Vega-Carrillo, H. R., Maldonado-Vega, M., Salas-Luévano, M. Á., & Argumedo-Delira, R. (2018). Induced accumulation of Au, Ag and Cu in *Brassica napus* grown in a mine tailings with the inoculation of *Aspergillus niger* and the application of two chemical compounds. *Ecotoxicology and Environmental Safety, 154,* 180–186. https://doi.org/10.1016/j.ecoenv.2018.02.055.

Harris, A. T., Naidoo, K., Nokes, J., Walker, T., & Orton, F. (2009). Indicative assessment of the feasibility of Ni and Au phytomining in Australia. *Journal of Cleaner Production, 17*(2), 194–200. https://doi.org/10.1016/j.jclepro.2008.04.011.

Harumain, Z. A. S., Parker, H. L., Muñoz García, A., Austin, M. J., McElroy, C. R., Hunt, A. J., Clark, J. H., Meech, J. A., Anderson, C. W. N., Ciacci, L., Graedel, T. E., Bruce, N. C., & Rylott, E. L. (2017). Toward financially viable phytoextraction and production of plant-based palladium catalysts. *Environmental Science and Technology, 51*(5), 2992–3000. https://doi.org/10.1021/acs.est.6b04821.

Haverkamp, R. G., Marshall, A. T., & Van Agterveld, D. (2007). Pick your carats: Nanoparticles of gold-silver-copper alloy produced in vivo. *Journal of Nanoparticle Research, 9*(4), 697–700. https://doi.org/10.1007/s11051-006-9198-y.

Heilmeier, H., & Wiche, O. (2020). The PCA of phytomining: principles, challenges and achievements. *Carpathian Journal of Earth and Environmental Sciences, 15,* 37–42. https://doi.org/10.26471/cjees/2020/015/106.

Hunt, A. J., Anderson, C. W. N., Bruce, N., García, A. M., Graedel, T. E., Hodson, M., Meech, J. A., Nassar, N. T., Parker, H. L., Rylott, E. L., Sotiriou, K., Zhang, Q., & Clark, J. H. (2014). Phytoextraction as a tool for green chemistry. *Green Processing and Synthesis, 3*(1), 3–22. https://doi.org/10.1515/gps-2013-0103.

Jaffré, T., Reeves, R. D., Baker, A. J. M., Schat, H., & Van Der Ent, A. (2018). The discovery of nickel hyperaccumulation in the New Caledonian tree *Pycnandra acuminata* 40 years on: an introduction to a Virtual Issue. *New Phytologist, 218,* 397–400. https://doi.org/10.1111/nph.15105.

Kidd, P. S., Bani, A., Benizri, E., Gonnelli, C., Hazotte, C., Kisser, J., Konstantinou, M., Kuppens, T., Kyrkas, D., Laubie, B., Malina, R., Morel, J. L., Olcay, H., Pardo, T., Pons, M. N., Prieto-Fernández, Á., Puschenreiter, M., Quintela-Sabarís, C., Ridard, C., & Echevarria, G. (2018). Developing sustainable agromining systems in agricultural ultramafic soils for nickel recovery. *Frontiers in Environmental Science, 6*(June), 44. https://doi.org/10.3389/fenvs.2018.00044.

Koptsik, G. N. (2014). Problems and prospects concerning the phytoremediation of heavy metal polluted soils: A review. *Eurasian Soil Science, 47*(9), 923–939. https://doi.org/10.1134/S1064229314090075.

Krisnayanti, B., Anderson, C., Sukartono, S., Afandi, Y., Suheri, H., & Ekawanti, A. (2016). Phytomining for artisanal gold mine tailings management. *Minerals, 6*(3), 84. https://doi.org/10.3390/min6030084.

Kumar, P. (2010). *The Economics of Ecosystems and Biodiversity: Ecological and Economic Foundations.* UNEP/Earthprint, London. ISBN: 9781849712125.

Li, C., Ji, X., & Luo, X. (2020). Visualizing hotspots and future trends in phytomining research through scientometrics. *Sustainability, 12*(11), 4593. https://doi.org/10.3390/su12114593.

Li, Y. M., Chaney, R., Brewer, E., Roseberg, R., Angle, J. S., Baker, A., Reeves, R., & Nelkin, J. (2003). Development of a technology for commercial phytoextraction of nickel: Economic and technical considerations. *Plant and Soil, 249*(1), 107–115. https://doi.org/10.1023/A:1022527330401.

Matko Stamenković, U., Andrejić, G., Mihailović, N., & Šinžar-Sekulić, J. (2017). Hyperaccumulation of Ni by *Alyssum murale* Waldst. & Kit. from ultramafics in Bosnia and Herzegovina. *Applied Ecology and Environmental Research, 15*(3), 359–372. https://doi.org/10.15666/aeer/1503_359372.

Midula, P., Wiche, O., Wiese, P., & Andráš, P. (2017). Concentration and bioavailability of toxic trace elements, germanium, and rare earth elements in contaminated areas of the Davidschacht dump-field in Freiberg (Saxony). *Freiberg Ecology Online, 2,* 101–112.

Naila, A., Meerdink, G., Jayasena, V., Sulaiman, A. Z., Ajit, A. B., & Berta, G. (2019). A review on global metal accumulators—Mechanism, enhancement, commercial application, and research trend. *Environmental Science and Pollution Research, 26,* 26449–26471. https://doi.org/10.1007/s11356-019-05992-4.

Nicks, L. J., & Chambers, M. F. (1995). Farming for metals. *Mining Environmental Management, 3*(3), 15–16.

Nicks, L. J., & Chambers, M. F. (1998). Pioneering study of the potential of phytomining for nickel. In: Brooks, R. R. ed. *Plants that Hyperaccumulate Heavy Metals: Their Role in Phytoremediation, Microbiology, Archaeology, Mineral Exploration and Phytomining,* 313–325. CAB International, Wallingford. ISBN: 9780851992365.

Nkrumah, P. N., Baker, A. J. M., Chaney, R. L., Erskine, P. D., Echevarria, G., Morel, J. L., & van der Ent, A. (2016). Current status and challenges in developing nickel phytomining: an agronomic perspective. *Plant and Soil, 406*(1–2), 55–69. https://doi.org/10.1007/s11104-016-2859-4.

Nkrumah, P. N., Chaney, R. L., & Morel, J. L. (2018). Agronomy of 'metal crops' used in agromining. In: Van der Ent, A., Echevarria, G., Baker, A. J. M., Morel, J. L. eds. *Agromining: Farming for Metals*, 19–38. Springer, Cham. https://doi.org/10.1007/978-3-319-61899-9_2.

Novo, L. A. B., Mahler, C. F., & González, L. (2015). Plants to harvest rhenium: Scientific and economic viability. *Environmental Chemistry Letters, 13*(4), 439–445. https://doi.org/10.1007/s10311-015-0517-3.

PhytoSUDOE. (2019). http://www.phytosudoe.eu/en/ (last accessed October 29, 2020).

Pilon-Smits, E. (2005). Phytoremediation. *Annual Review of Plant Biology, 56*(1), 15–39. https://doi.org/10.1146/annurev.arplant.56.032604.144214.

Rascio, N., & Navari-Izzo, F. (2011). Heavy metal hyperaccumulating plants: How and why do they do it? And what makes them so interesting? *Plant Science, 180*(2), 169–181. https://doi.org/10.1016/j.plantsci.2010.08.016.

Rentsch, L., Aubel, I. A., Schreiter, N., Höck, M., & Bertau, M. (2016). PhytoGerm: Extraction of germanium from biomass – An economic pre-feasibility study. *Journal of Business Chemistry, 13*(1), 47–58.

Robinson, B., Fernández, J. E., Madejón, P., Marañón, T., Murillo, J. M., Green, S., & Clothier, B. (2003). Phytoextraction: An assessment of biogeochemical and economic viability. *Plant and Soil, 249*(1), 117–125. https://doi.org/10.1023/A:1022586524971.

Robinson, B. H., Brooks, R. R., Howes, A. W., Kirkman, J. H., & Gregg, P. E. H. (1997a). The potential of the high-biomass nickel hyperaccumulator *Berkheya coddii* for phytoremediation and phytomining. *Journal of Geochemical Exploration, 60*(2), 115–126. https://doi.org/10.1016/S0375-6742(97)00036-8.

Robinson, B. H., Chiarucci, A., Brooks, R. R., Petit, D., Kirkman, J. H., Gregg, P. E. H., & De Dominicis, V. (1997b). The nickel hyperaccumulator plant *Alyssum bertolonii* as a potential agent for phytoremediation and phytomining of nickel. *Journal of Geochemical Exploration, 59*(2), 75–86. https://doi.org/10.1016/S0375-6742(97)00010-1.

Robinson, B. H., Bañuelos, G., Conesa, H. M., Evangelou, M. W. H., & Schulin, R. (2009). The phytomanagement of trace elements in soil. *Critical Reviews in Plant Sciences, 28*(4), 240–266. https://doi.org/10.1080/07352680903035424.

Rodrigues, J., Houzelot, V., Ferrari, F., Echevarria, G., Laubie, B., Morel, J. L., Simonnot, M. O., & Pons, M. N. (2016). Life cycle assessment of agromining chain highlights role of erosion control and bioenergy. *Journal of Cleaner Production, 139*, 770–778. https://doi.org/10.1016/j.jclepro.2016.08.110.

Rosenkranz, T., Hipfinger, C., Ridard, C., & Puschenreiter, M. (2019). A nickel phytomining field trial using *Odontarrhena chalcidica* and *Noccaea goesingensis* on an Austrian serpentine soil. *Journal of Environmental Management, 242*, 522–528. https://doi.org/10.1016/j.jenvman.2019.04.073.

Rosenkranz, T., Kisser, J., Wenzel, W. W., & Puschenreiter, M. (2017). Waste or substrate for metal hyperaccumulating plants — The potential of phytomining on waste incineration bottom ash. *Science of the Total Environment, 575*, 910–918. https://doi.org/10.1016/j.scitotenv.2016.09.144.

Saxena, G., Purchase, D., Mulla, S. I., Saratale, G. D., & Bharagava, R. N. (2020). Phytoremediation of heavy metal-contaminated sites: Eco-environmental concerns, field studies, sustainability issues, and future prospects. *Reviews of Environmental Contamination and Toxicology, 249*, 71–131. https://doi.org/10.1007/398_2019_24.

Sheoran, V., Sheoran, A. S., & Poonia, P. (2009). Phytomining: A review. *Minerals Engineering, 22*(12), 1007–1019. https://doi.org/10.1016/j.mineng.2009.04.001.

Sheoran, V., Sheoran, A. S., & Poonia, P. (2013). Phytomining of gold: A review. *Journal of Geochemical Exploration, 128,* 42–50. https://doi.org/10.1016/j.gexplo.2013.01.008.

Shi, P., Veiga, M., & Anderson, C. (2020). Geochemical assessment of platinum group metals for phytomining. *Revista Escola de Minas, 73*(1), 85–91. https://doi.org/10.1590/0370-44672019730038.

Simonnot, M. O., Laubie, B., Zhang, X., Houzelot, V., Ferrari, F., Rodrigues, J., Pons, M. N., Bani, A., Echevarria, G., & Morel, J. L. (2016). *Agromining: Producing Ni salts from the biomass of hyperaccumulator plants.* IMPC 2016 – 28th International Mineral Processing Congress, September 2016.

Simonnot, M.-O., Vaughan, J., & Laubie, B. (2018). Processing of bio-ore to products. In: Van der Ent, A., Echevarria, G., Baker, A. J. M., and Morel, J. L. eds. *Agromining: Farming for Metals,* 39–51. Springer, Cham. https://doi.org/10.1007/978-3-319-61899-9_3.

Tang, Y. T., Deng, T. H. B., Wu, Q. H., Wang, S. Z., Qiu, R. L., Wei, Z. Bin, Guo, X. F., Wu, Q. T., Lei, M., Chen, T. Bin, Echevarria, G., Sterckeman, T., Simonnot, M. O., & Morel, J. L. (2012). Designing cropping systems for metal-contaminated sites: A review. *Pedosphere, 22*(4), 470–488. https://doi.org/10.1016/S1002-0160(12)60032-0.

Van der Ent, A., Baker, A. J. M., Reeves, R. D., Chaney, R. L., Anderson, C. W. N., Meech, J. A., Erskine, P. D., Simonnot, M. O., Vaughan, J., Morel, J. L., Echevarria, G., Fogliani, B., Rongliang, Q., & Mulligan, D. R. (2015). Agromining: Farming for metals in the future? *Environmental Science and Technology, 49*(8), 4773–4780. https://doi.org/10.1021/es506031u.

Van der Ent, A., Baker, A. J. M., van Balgooy, M. M. J., & Tjoa, A. (2013). Ultramafic nickel laterites in Indonesia (Sulawesi, Halmahera): Mining, nickel hyperaccumulators and opportunities for phytomining. *Journal of Geochemical Exploration, 128,* 72–79. https://doi.org/10.1016/j.gexplo.2013.01.009.

Wang, L., Hou, D., Shen, Z., Zhu, J., Jia, X., Ok, Y. S., Tack, F. M. G., & Rinklebe, J. (2020). Field trials of phytomining and phytoremediation: A critical review of influencing factors and effects of additives. *Critical Reviews in Environmental Science and Technology, 50*(24), 2724–2774. https://doi.org/10.1080/10643389.2019.1705724.

Wiche, O., & Heilmeier, H. (2016). Germanium (Ge) and rare earth element (REE) accumulation in selected energy crops cultivated on two different soils. *Minerals Engineering, 92,* 208–215. https://doi.org/10.1016/j.mineng.2016.03.023.

Wiche, O., Székely, B., Kummer, N. A., Moschner, C., & Heilmeier, H. (2016). Effects of intercropping of oat (*Avena sativa* L.) with white lupin (*Lupinus albus* L.) on the mobility of target elements for phytoremediation and phytomining in soil solution. *International Journal of Phytoremediation, 18*(9), 900–907. https://doi.org/10.1080/15226514.2016.1156635.

Wiche, O., Tischler, D., Fauser, C., Lodemann, J., & Heilmeier, H. (2017). Effects of citric acid and the siderophore desferrioxamine B (DFO-B) on the mobility of germanium and rare earth elements in soil and uptake in *Phalaris arundinacea.* *International Journal of Phytoremediation, 19*(8), 746–754. https://doi.org/10.1080/15226514.2017.1284752.

Willscher, S., Mirgorodsky, D., Jablonski, L., Ollivier, D., Merten, D., Büchel, G., Wittig, J., & Werner, P. (2013). Field scale phytoremediation experiments on a heavy metal and uranium contaminated site, and further utilization of the plant residues. *Hydrometallurgy, 131–132,* 46–53. https://doi.org/10.1016/j. hydromet.2012.08.012.

Wilson-Corral, V., Anderson, C. W. N., & Rodriguez-Lopez, M. (2012). Gold phytomining. A review of the relevance of this technology to mineral extraction in the 21st century. *Journal of Environmental Management, 111,* 249–257. https://doi. org/10.1016/j.jenvman.2012.07.037.

Wilson-Corral, V., Anderson, C. W. N., Rodriguez-Lopez, M., Arenas-Vargas, M., & Lopez-Perez, J. (2011). Phytoextraction of gold and copper from mine tailings with *Helianthus annuus* L. and *Kalanchoe serrata* L. *Minerals Engineering, 24*(13), 1488–1494. https://doi.org/10.1016/j.mineng.2011.07.014.

5

Establishing Miscanthus, Production of Biomass, and Application to Contaminated Sites

Lawrence C. Davis, Valentina Pidlisnyuk, Aigerim Mamirova, Pavlo Shapoval, and Tatyana Stefanovska

Abstract

The establishment of vegetation on sites with contamination to improve soil quality, reduce risk, and produce a biomass product depends on many local conditions. Site characterization results, remediation goals, local markets for biomass products, soil properties, climate, temperature, annual and seasonal precipitation, past experience at similar sites, and availability of soil amendments are some aspects to consider. This chapter addresses the establishment of Miscanthus and its application when growing in contaminated soil. Plant selection and breeding of Miscanthus is reviewed briefly. Water is very important for the establishment of Miscanthus, and issues related to rainfall during the first weeks and months are reviewed. Plant nutrition and soil amendments affect growth and biomass yield. The time of harvest for Miscanthus affects nitrogen use because much more nitrogen is removed if the harvest is in the fall compared to winter harvest. The effects of soil amendments on the fate of contaminants and plant growth are included because the knowledge of how to obtain beneficial results by adding soil amendments has advanced significantly. Improved results have been reported for Miscanthus production when plant growth regulators have been added. Results from recent literature are included.

CONTENTS

5.1 Plant Selection and Breeding

The cultivar (CV) of *Miscanthus × giganteus (M. × giganteus)* that is used in most studies in the US and many places in Europe is a sterile triploid hybrid of diploid *M. sinensis* (receiving 1n) and tetraploid *M. sacchariflorus* (receiving 2n) but with some additional ancestral contribution from *M. sinensis* (Mitros et al., 2020). It was brought from Japan to Europe by 1930 and sold as a decorative plant until the 1980s when it was promoted, initially in Denmark, as a potential source of biomass (Nielsen, 1990; Pude, 2008). The most widely grown clone in the US was derived from material at the Chicago Botanical Garden and later designated as the Illinois clone. A CV named "Freedom" developed at University of Mississippi from a USDA germplasm collection appears to have essentially the same genotype. A strain obtained by K-State from Bluemel Nursery in 2007 is actually tetraploid *M. sacchariflorus*. Below it is identified as *M. sacchariflorus* Bluemel. Material obtained from Bluemel in 1987 and grown at the Minnesota Arboretum in Minneapolis appears identical in observable phenotype, to the Illinois clone of *M. × giganteus* obtained from Maple River Farms of Owosso, Michigan in 2018 (Davis L. personal observation). The Bluemel Nursery website states that Bluemel obtained his original *M. × giganteus* material in Europe (Switzerland) in 1960. Thus, we are working with a very narrow genetic base and there is no reason to think that we have a best widely adapted hybrid.

Some groups have attempted to improve performance by making triploid or tetraploid hybrids of the two above-mentioned species, or by development of selections from either of the parental species for desirable traits. The parental species have relatively wide distributions within East Asia and one can find a wide range of phenotypes with different winter hardiness, daylength sensitivity of flowering, and other important characteristics. Kalinina et al. (2017) reported in detail a program to test multiple germplasms, plus some new crosses, with *M. ×giganteus* for a reference at multiple sites over a wide range of climatic conditions. Results varied greatly over both years of study, and sites, across Europe from Wales to Russia and Turkey. One triploid hybrid *M. ×giganteus*, "Nagara" bred in Europe by M. Deuter, has been used to some extent in North America but its expected winter hardiness, observed in Urbana, IL (Dong et al., 2019), has not proved out in practice in Ontario, Canada (Sage et al., 2015).

There are several major practical challenges with managing Miscanthus hybrids. One is concern for potential invasiveness of new fertile lines. Self-incompatibility may reduce problems with a single clone, but it is a hazard of seed-grown hybrids. Both of the parental species of *M. ×giganteus* are classified as invasive in some states within the US. That does not mean that they are prohibited, but it reduces their acceptability for widespread distribution and growth, because they may become seriously invasive under some climatic conditions.

Further, in the US there are not mandates for use of carbon neutral strategies unlike in Europe (Clifton-Brown et al., 2017) so that the biofuels market is not good. Ethanol from maize grain is the largest such biofuel and cellulosic, derived from fermentation of cellulose, are unable to compete economically. There is a small market for the direct use of *M. ×giganteus* as fuel, animal bedding, or dietary fiber, but that is satisfied by already available *M. ×giganteus* materials (Moberly Monitor, 2017; UIFM, 2020). The situation is quite different in Europe, so that some large breeding programs continue there.

An excellent review of the longest continuous breeding program, based in Wales, is described by Clifton-Brown et al. (2019). Four tracks of breeding strategy were used to make most rapid progress with a perennial crop, with a goal of identifying a seed-based production plan, rather than clonal propagation as used with *M. ×giganteus*. High yield, cold tolerance, drought tolerance, climatic adaptability over a wide geographic area were among the many characteristics being searched for in progeny of thousands of possible crosses from a starting point of 240 genotypes.

A large practical concern in breeding and selection is for propagation and establishment from seed (Xue et al., 2015). Miscanthus seeds are quite small and the plantlets do not become competitive with agricultural weeds until after at least a season's growth. They are also very sensitive to drought immediately after planting, or in the following months (Anderson et al., 2015). These factors mean that any seed-propagated crop will require intensive care in the first year. This makes it uncompetitive with even the costly propagation of clonal *M.* × *giganteus*, for which methods of propagation have been improved. A single improved clone such as Nagara, which is also a sterile triploid, might be competitive if it has a significant advantage over *M.* × *giganteus* in yield, hardiness, or ease of propagation.

5.2 Plant Establishment

For *M.* × *giganteus*, good establishment of a productive field can be challenging. It reaches peak production after 2–4 years and may last in excess of two decades, so that doing it right the first time is expected to pay off. It is not inexpensive to do correctly.

Four critical factors for success are weight (of the propagule); water (to promote root growth); weeds (which must be controlled); weather (which cannot be controlled). Anderson et al. (2011) provide a good example for one specific region of the US where production of biomass approaches its maximum.

5.2.1 Weight

It has been observed that for direct planting in a well tilled field, likelihood of success in establishment increases with propagule weight. Typically, dormant rhizomes are purchased/provided with a weight of less than 25–30 g. Success is much higher when they weigh at least 60 g (Pyter et al., 2010). Preparation costs per propagule, which includes number that can be obtained per unit area of nursery, amount of material to be dug, sorted, and packed, and shipping weight charges, become issues as mass per propagule increases (Figure 5.1).

An alternative which in some settings can reduce costs is using stem buds for propagation in a winter nursery in a warm climate/greenhouse, and then shipment of the actively growing plants, commonly called plugs, to the planting sites. This has been successful in Iowa, USA (Boersma & Heaton, 2012, 2014a, b). For large-scale biomass production the University of Iowa contracted with a commercial plant propagation company to grow small individual plants during winter, prepared to ship at the appropriate planting time in late April. Each stem from a rhizome can yield about five bud plants, from the most basal, commonly underground, buds, and each 25 g propagule may yield two shoots which allows a ten-fold more rapid increase of plants than simply using rhizome pieces from a second or third year field

(a) (b)

FIGURE 5.1
(a) Three sizes of rhizome piece used for propagation, with fresh weights of about 10, 25, and 50 g. The number of active shoot buds and total reserves is approximately proportional to size. Terminal buds are most active, a kind of apical dominance. The 10 g pieces are suitable for planting in an intensively managed nursery to produce larger propagules for the next year. The 25 g piece is of a size typically used for planting. The 50 g piece is at the minimum of preferred size for a high percentage of successful establishment. It possesses multiple potential shoots at rhizome tips. (b) An ideal propagule, found at the base of a 10 ft tiller, with three distinct rhizomes attached, each of which will grow one strong tiller in the following spring and total weight >60 g. Such good materials are found at the perimeter of 2- and 3-year-old clumps. Older interior rhizomes are less effective, less vigorous.

of *M. ×giganteus* (Boersma and Heaton, 2014a, b). However, often the costs of this mode of stem and rhizome bud propagation exceed that of using rhizomes directly. Micropropagation was found by Kołodziej et al. (2016) to be more expensive in southeast Poland than use of rhizomes, in part because of poor plantlet survival in field conditions (~80%) and the need for replacement plants, but also because the price of rhizomes is relatively lower by a factor of 2–3. For new CVs, in vitro micropropagation may be worthwhile but it is costlier than either of the above techniques according to Kołodziej et al. (2016).

Comparing rhizome to plantlet propagation with *M. ×giganteus*, Ouattara et al. (2020) tested at six diverse sites in northern and central France followed over 6 years. They found that the rhizomes were less effective for both establishments in year 1 and regrowth in year 2. Establishment was 77% and regrowth was 86% compared to 87% and 94% for the plantlets from small rhizomes (~10 g) or stem buds. By comparison seed-grown *M. sinensis* (not directly field seeded but started as plantlets, mixed genotypes) gave high establishment and regrowth, the same as *M. ×giganteus* plantlets. *M. sinensis* had lower yields than *M. ×giganteus* over all years, but less year to year variability.

A significant hindrance to direct sowing of Miscanthus seed is allelopathy. Awty-Carroll et al. (2020) documented this problem and identified a complex mixture of extractables present in seed including proanthocyanidins and vanillic acid. This makes it difficult to use seed clusters to enhance stands. In controlled competition there appeared to be root competition and it was also observed in field scale that shoot competition was a problem. Thus, when using direct seeding, it proved a challenge to obtain good stands.

The relative price and success of rhizome vs plantlet propagation seem to differ between situations and countries. Much of the variation in success may depend on summer water and winter weather. Overplanting by 25%–50% may be an economically viable alternative to assure an adequate stand (Boersma & Heaton, 2014a).

5.2.2 Water

Adequate water is essential to establish growth of rhizomes. Available water is decreasing in many areas while being found in excess in others. Caslin et al. (2011) report that soil moisture must be at least 40% of field capacity at planting. The rhizomes desiccate rapidly when kept at ambient temperature without water. Drought at this stage is fatal. If rain is insufficient, supplemental water is necessary. Providing supplemental water to 1 ha of *M. ×giganteus* (>10,000 plants) can be costly. Drip irrigation would require 1.1 km of drip tubing with nozzles at plant spacing (up to 1 m apart), plus additional fittings. With drip irrigation, watering needed will depend on the soil water-holding capacity within a zone comprising several liters surrounding each rhizome. The equivalent of 10 cm depth of water in an area 15 cm in radius (7 L) will wet soil to a depth of about 30 cm. This uses 70 m^3 of water for 10,000 plants. To supply each plant with a supplemental 10 cm water uses 1000 m^3 of water if using sprinkler irrigation.

For a first-year crop, drip irrigation can allow large savings if water is expensive. During peak growth later in the establishment year, water use may be greater than 5 cm depth wk^{-1}. There must be sufficient supply to have significant percolation of water to depths greater than the root depth during establishment, to encourage their downward growth. Roots of *M. ×giganteus* are reported to reach depths of >2 m, though the large majority of roots, when water is available, remain much closer to the surface. Rhizomes of *M. ×giganteus* are usually within the top 15 cm (Sage et al., 2015) (Figure 5.2).

In the study described by Kalinina et al. (2017), drip irrigation was used in Turkey during second to fourth years, with sufficient irrigation (23 cm in year 2) added to nearly match the potential evapotranspiration. The amount of irrigation used in year 1 was not stated but it was enough to result in rapid growth of the plants, so that yields, averaged over all trialed CVs that were better in the first year than other sites, were also better in the second year (>10 t ha^{-1}), though not for lack of water in the more temperate settings of other sites. The longer, warmer growing season in Turkey may have been largely responsible for higher production. See discussion below about geography of factors for Miscanthus growth.

FIGURE 5.2
Excavated section at the edge of a plot of *M. ×giganteus* showing tillers arising from ground with prominent rhizomes, long thin exploratory roots, and short highly branched roots that explore the soil for water and nutrients. The length of roots penetrating into deeper soil depends on water availability and soil porosity. The green tiller at center right is shown as Figure 5.1b, a perfect propagule.

When the canopy closes and the crop matures to optimum production, after a minimum of 2–3 years but sometimes near a decade, water use is at least 200 L kg^{-1} aboveground dry matter accumulation (Clifton-Brown & Lewandowski, 2000; Mantineo et al., 2009). In dry climates, more water is lost than in humid ones. Hot and windy conditions also increase water losses. The highest reported total dry matter production may reach 40 t ha^{-1} which requires a minimum of 0.8 m water during the active growing season. This usually implies a total rainfall or irrigation of >1 m year^{-1}. Moisture deficit explained 70% of the variance in *M. ×giganteus* yield, at Rothamstead, UK, which had the largest available data set (Richter et al., 2008) with yields that ranged from 5 to 18 t ha^{-1}. There was an average of 12.8 t ha^{-1} across 14 UK sites over 3 years.

There are few quantitative reports on minimum water requirements for successful growth of *M. ×giganteus* in relation to different climates. In southern Oklahoma (34.2 N latitude) yields are consistently low because summer rainfall is insufficient (Kering et al., 2012). Temperatures are high, probably too high, and skies are clear. Fully irrigated maize grown somewhat further north in southwest Kansas (~38 N latitude), with some of the highest solar availability in the nation, produces yields to 19 t ha^{-1} grain and near 38 t ha^{-1} total dry matter (KCYCW, 2019). Comparable maize grain yields are obtained in southern Nebraska (~41 N latitude) at over 17 t ha^{-1}, near 34 t ha^{-1} biomass, with contest winners exceeding 19 t ha^{-1} of grain (NCGA, 2019).

Miscanthus yields may be higher in temperate zones of southern Europe, e.g., Croatia northward (~45 N latitude), where longer daylength promotes summer growth and delays flowering of *M. × giganteus* if water is not limiting. On the other hand, yields of *M. × giganteus* in the US are not generally higher in southern states, below ~30 N latitude, where daylength is shorter sooner, even when rainfall and soil fertility are not limiting. Lee et al. (2018) provide results for maintenance stages of *M. × giganteus* production in comparison with other biomass energy crops (selected cultivars of switchgrass, sorghum, and sugarcane) in the U.S. Rainfall, and in some cases fertilizer application, affected productivity at most sites, after successful establishment. Productivity was most variable in Mead Nebraska (~41.2 N latitude), ranging from 15 to 35 Mg ha^{-1}, while other sites had intermediate but more stable yields. This was primarily a function of variable rainfall in Nebraska plus a variable long growing season.

Lee et al. (2017) reported a strong interaction of water availability and N fertilization at Urbana, IL (~40 N latitude). For an established study (from 2008 to 2009) yields declined in a droughted summer of 2012 for the unfertilized plots while remaining relatively high and stable with N fertilization at 60 kg ha^{-1} year^{-1}. Plots were 10×10 m and 100 plants, with four replicates of each treatment. Water availability was extremely reduced, 68% and 90% below historic averages over 2 months, June and July in 2012, August and September in 2013. Yields for 0 kg ha^{-1} added N vs 60 kg ha^{-1} were: in 2011, 15.9 vs 23.3 Mg kg ha^{-1}; in 2012, 11.6 vs 24.8; in 2013, 15.3 vs 28.8; in 2014, 8.5 vs 25.9. The low rainfall in 2013 continued throughout the remainder of the year at <1/2 of the 30-year average (28 vs 61 cm) for the latter half of the year. Although rainfall in 2014 was above average, the unfertilized plot may have been unable to take advantage of that abundance. Averaged over the 4 years, yield in the fertilized plots was double that of the unfertilized (25.5 vs 12.8 Mg ha^{-1} year^{-1}).

It should be noted, however, that Lee et al. (2018) reported rather different yield data for experiments with the same treatment design at the same, or a very nearby location. For instance, in 2014 the unfertilized plots yielded the same as in 2013, while in 2012 yields of both unfertilized and fertilized plots were about half of their respective yields in 2013. Neither paper offers any comment on the large apparent differences. Sampling of biomass yield seems to have been done differently in the two studies. Lee et al. (2017) used five 1 m^2 quadrats per plot while Lee et al. (2018) used one 4 m^2 sample at the center of the 10×10 m plot. Even with such a large sample, some replicates in some years varied from others within a treatment, by more than two-fold as shown in their supplemental data. These results indicate that interpreting Miscanthus yields in terms of known variables such as rainfall is challenging.

In a dry summer Mediterranean climate (Sicily) restoring 25% (~15 cm) or 75% (~45 cm) potential evapotranspiration water by irrigation increased yields of both *Arundo donax* and *M. × giganteus* energy crops studied by

Mantineo et al. (2009). The common name of *Arundo donax* is giant cane or giant reed, and it bears some similarity to *M. ×giganteus* in adaptability and productivity, as discussed below. Even when irrigation was stopped in the fourth year of study, *M. ×giganteus* yielded over 27 Mg ha^{-1}. Yield decreased somewhat in the fifth year with no irrigation. Significant winter rainfall must have been stored in the soil column, although soil water capacity was reported as ~10 cm in the top 0.8 m of total 1.2 m soil depth.

Successful *M. ×giganteus* establishment in this Mediterranean climate was absolutely dependent on irrigation because of very limited summer rainfall (Mantineo et al., 2009). Cosentino et al. (2007) had earlier measured Water Use Efficiency (WUE) for *M. ×giganteus* grown with limited irrigation in the same environment and found it was above 4.5–4.8 g L^{-1} aboveground dry matter production in consecutive years. They noted that the WUE declined as more water was provided to ~2.5 and 3.5 g L^{-1}. The reported numbers are consistent with the work of Clifton-Brown & Lewandowski (2000) in a greenhouse experiment where water amounts are more easily measured. The work of Cosentino et al. (2007) and Mantineo et al. (2009) suggests that *M. ×giganteus* can produce high yields with less water than most crops.

In no year does the *M. ×giganteus* WUE reach the WUE of *A. donax* in the same location, which was above 4 g L^{-1} beginning with the second year. The authors note that *A. donax* did not go totally dormant in winter, unlike *M. ×giganteus*, and so may it make better use of available water in winter. Of course, in colder climates such advantage is not available. The aboveground vegetation of *A. donax* like that of *M. ×giganteus* is frost sensitive, though perhaps not so much so. Below-ground rhizomes are killed at about −5°C for 24 hours, a bit more tolerant than *M. ×giganteus* (Pompeiano et al., 2015).

It is clear from the above studies, and general knowledge of crop physiology, that water-holding capacity of soil, in addition to total applied water, and deep roots are all critical for maximizing yield when dependent on rainfall. One recent study from Ukraine (Doronin et al., 2019) shows that the use of water-retaining materials can benefit rhizome development when that is the goal, as in the preparation of large rhizomes for transplanting. Water-holding materials can allow for longer survival in transitory droughts and are widely used in potting materials for planters that are intermittently watered. This might not be economical for field-scale production but could benefit nursery preparation of planting materials.

5.2.3 Weeds

Detailed studies of herbicide tolerance of *M. ×giganteus* were done by Eric Anderson for his M.S. degree at University of Illinois (2010). Tolerance to broad-leaf herbicides (mainly auxin analogs) was the same as for maize, as expected. Tolerance of wider-spectrum herbicides and of those designed for control of grassy weeds were also reported. A more extensive greenhouse study by Smith et al. (2015a) considered 22 PRE- and 22 POST-planting herbicides with several energy crops. They noted that *M. ×giganteus* from rhizomes was more tolerant than seedlings of hybrid Miscanthus.

Careful field preparation and use of a broad spectrum, short-lived herbicide such as glyphosate has been found to be important for obtaining good vigorous *M. ×giganteus* stands in the US and UK (Caslin et al., 2011; USDA/NRCS, 2011). For regulatory reasons we cannot make specific recommendations for U.S. registered herbicides in Europe or Asia, and vice versa. Song et al. (2016) confirm the essential need for strong weed control and provide a useful example of the range of herbicides that are effective with *M. sacchariflorus* in Korea. Similar results are described for *M. ×giganteus* (Roik et al., 2019).

Management of weeds after *M. ×giganteus* has begun growing requires special care to not damage the plants. Tillage methods can only be used between rows, carefully avoiding close approach to the plants themselves. Herbicide application must be under strictly defined conditions of temperature, soil moisture, and wind speed to meet regulatory requirements and avoid damaging stressed plants. Because *M. ×giganteus* is a single clone; no genetically modified (GM) form of the plant is available to make use of resistance to specific herbicides, unlike maize resistant to glyphosate and dicamba.

A comparison study in Kentucky and Virginia by Smith et al. (2015b) showed that weed competition in those two locations, on ground previously in no-till rotations with winter cover-cropping, was not a significant obstacle for several energy crops including *M. ×giganteus* CVs Illinois, Nagara, and (tetraploid fertile) PowerCane. The main difference between this study and the earlier work in Illinois was that the soils of Illinois are richer in N, having been in crop rotations requiring high N levels, mainly maize. They may also have larger seed banks of weedy species.

Our first effort to grow Miscanthus in the field began May 1, 2015, at Ft. Riley, KS (~39 N latitude). It made use of large propagules of the *M. ×giganteus* Bluemel strain kept 1 month or 13 months at 4°C, and smaller plantlets grown under lights during the previous summer, autumn, and winter. Those were grown under several conditions including soil from the planting site, with or without fertilizer; perlite, or perlite plus vermiculite both with 1/2 strength Hoagland's solution. Rhizomes (60–90 g) harvested from larger potted (~12 L) plants stored cold 6 or 18 months, and freshly harvested (from a field site) actively growing rhizomes were also tested. Some plantlets were grown up and then stored 5 months in cold.

The site was grassland, with alfalfa, normally mowed several times per year. All vegetation was cut to ~1–2 cm with a "weedeater" and propagules were planted at 45 cm intervals in 11 rows, spaced 45 cm apart. Including border edges, the plot was 6 × 6 m. Fertilizer, tillage, and herbicide treatments were avoided. Every Miscanthus plant was marked with a bamboo stick. Rain was abundant, with three intervals of >10 cm rain, after 3 days, during the following 3 weeks and again during 2 weeks. Weeds were trimmed down to

<10 cm on two occasions, June 1 and June 18. Rain was adequate through the remainder of the season. No further weed control was used.

Survival was recorded on May 5, 2016, when plants had emerged from winter dormancy. The plantlets stored cold in perlite or perlite + vermiculite and knocked off at planting, had as fraction of survivors, 7/11; with mixture left on, 8/11; small rhizome propagules, or actively growing rhizomes 9/11; all other treatments 11/11. Some of the more vigorous plants in one border row had extended runner rhizomes over 50 cm out into the surrounding area. In December 2016 this plot had a biomass of 8.0 ± 1.9 t ha^{-1} (N = three samples of 0.81 m^2) comparable to the 2016 year 1 yield of the study by Alasmary (2020), which was done immediately adjacent to this trial plot.

5.2.4 Weather

While climate change is a long-term transition that may alter the optimum growth regions for *M. ×giganteus*, changeable weather is the present challenge. Average winters have become warmer over much of the US, but occasional sudden changes are much more damaging to *M. ×giganteus*. Late spring frosts or sudden early autumn frosts can have large consequences (Kaiser and Sacks, 2015). In springtime, energy expended in sending forth shoots is lost when a sudden freeze severely injures them.

A major collaborative study of Miscanthus production and efforts to model its productivity in the US (Lee et al., 2018) has 58 authors. It was initiated in 2008 and compared Miscanthus with several other potential bioenergy sources including switchgrass, forage sorghum, "energycane" which is a version of sugarcane, and mixed grasses. Weather and geography were obvious variables. The duration and locations of trials for the different crops varied. For Miscanthus a 6-year study was done in five locations. At two initial locations, in Illinois and Indiana, there was high winter mortality of the transplants which had been started from ~25 g propagules in a greenhouse prior to planting in June/July 2008 (rather late). In Illinois 75% of plants were lost. The severely damaged Indiana site was replaced by one in the Virginia Piedmont in 2010. That site is climatically milder but with soils and terrain generally less suited to annual row crops.

We observed signs of significant prior frost damage on early emerging *M. sacchariflorus* Bluemel when we were planting on May 2, 2017, 2 weeks after the expected date of the last killing frost in our climate. The year-old plants in the study of Alasmary (2020) were ~50 cm tall by May 2, showing damaged foliage. A 2-year-old (2015) trial plot which had not been harvested the previous autumn had much shorter plants with far less damage (personal observation). Residual vegetation presumably delayed soil warming and shoot emergence.

Sudden fall frosts are reported to damage first year *M. ×giganteus* in Illinois because the plants tend to remain green longer during their first year (Boersma et al., 2015). This frost damage in turn reduces the translocation of nutrients back to the rhizome (Aurangzaib, 2012). Sage et al. (2015) observed and quantified a similar effect on a new triploid *M. ×giganteus* hybrid (Nagara) in Ontario, Canada. In this instance 3/4 of the leaf N failed to translocate. Similar effects may be observed for other nutrients.

During winters, especially the first winter, *M. ×giganteus* is vulnerable to freezing damage. When soil temperature drops below −3°C to −5°C at a depth of 4–10 cm, rhizomes are killed (Dong et al., 2019; Heaton et al., 2010). Insulation of the planting with straw, dropped foliage, or other means may be critical for good establishment in places where such cold is likely. In a long-term study described by Maughan et al. (2012) and Lee et al. (2017), 75% of a 2008 summer planting of potted, actively growing rhizomes was lost in the first winter. These had been greenhouse grown from 25 g rhizomes in $9 \times 9 \times 12$ cm pots so they were much larger than typical plantlets, but were planted rather late in mid-July. The missing plants were replaced in 2009 (Maughan et al., 2012).

A fall planting of *M. ×giganteus* ~25 g rhizomes at Mimon, Czech Republic, was fully destroyed by an 8-day February–March period of air minimum temperatures at −10°C to −18°C and maxima below 0. Soil temperatures were not determined. Only 1/10 of a previous spring planting in that same location survived (Nebeska D. personal observation).

Kucharik et al. (2013) developed a predictive model for likelihood of winter loss, based on a 30-year climate record across the Midwest US. The effect of insulation of the crop by straw and leaf matter at various depths was also calculated. A layer of 5 cm gave significant protection in many regions but the northern portions of several states near the Canadian border (~49 N latitude) had a 50% likelihood of losses even with protection. So far as we are aware no comparable model has been generated for Europe, although it is feasible to do so. Somewhat surprisingly, in the studies described by Kalinina et al. (2017), *M. ×giganteus* survived winter even in the vicinity of Moscow, Russia (~55.75 N latitude), because soil temperatures never dropped below 0°C in the winter of 2012–2013.

Seasonal droughts of varying magnitude have a strong negative effect on biomass yield, when irrigation is unavailable. Kering et al. (2012) described this effect in southern Oklahoma where half of carefully pot-grown *M. ×giganteus* plants died during the 2 months of 2008 following transplantation to a field site, while other biomass grasses had high survival. Biomass yield of the *M. ×giganteus* never approached that of the other crops over the study period even though precipitation returned from 58 cm in 2008 to near average 97 cm year^{-1} value in the second and third years (130 and 91 cm, respectively). In Eastern Ukraine, a planting of *M. ×giganteus* failed in 2017 when summer rainfall dropped from an average of 7 to only 1 cm during July although the annual total was close to the average 45 cm (Stefanovska T. personal observation).

Failures of timely rain following a 2017 planting at Ft. Riley, Kansas (~39 N latitude), resulted in loss of ~1/4 of the *M. sacchariflorus* Bluemel

(Davis L. personal observation) contrasting with zero losses in 2016 with abundant rain (Alasmary, 2020). During 2018 a direct comparison was done with *M. ×giganteus* and the *M. sacchariflorus* Bluemel strains both at the Ft. Riley site and at the Kansas State University North Farm site (~39.2 N latitude). At the former location there was a failure of rains over several weeks, and nearly total loss (>90%) of both CVs (two replicates each of 16 plants planted on 45 cm centers). At the North Farm where some supplemental irrigation was available, four replicate plots each of 36 plants grown on 45 cm centers were compared. Survival of the Bluemel strain was much higher (~3/4) than that of *M. ×giganteus* (<1/2) perhaps because larger propagules of 60–90 g vs 25 g were available. Even in autumn 2020, the *M. ×giganteus* plants remain as individual clumps of <50 cm diameter, while the Bluemel strain plants, with no supplemental irrigation since early summer 2017, are invading the entire planting (Davis L. personal observation). This is consistent with observations mentioned above where it sometimes takes more than 4 years to reach maximum productivity which depends on total number of stems (tillers) and number of leaves on each (Lee et al., 2017) (Figures 5.3 and 5.4).

In southeastern Kansas the Mound Valley experiment station (~37.2 N latitude) had an average yield of 11.3 t ha^{-1} for the third to fifth years of growing Illinois and "Freedom" *M. ×giganteus* CVs (which are genetically near identical *M. ×giganteus*). There was no correlation of yield to annual rainfall amounts of 91, 130, and 104 cm in years 3–5 (2014–2016). In the establishment year (2012) a significant fraction of the plants failed despite occasional irrigation. These were replaced in 2013 and the 2013 yield was about 5.4 t ha^{-1}. In

FIGURE 5.3
Comparison of characteristic leaf form of *M. sacchariflorus* (lower right), with *M. ×giganteus* (upper left) on September 11, 2018 prior to flowering of the latter. Leaves of *M. sacchariflorus* are wider than those of *M. ×giganteus*.

FIGURE 5.4
Direct comparison of growth habit of *M. ×giganteus* (front right) and *M. sacchariflorus* Bluemel (left, and behind) at the North Farm site, K-State, KS, in the fall of 2018. The *M. ×giganteus* was obtained from Maple River Farms and planted on May 1, 2018. It produced flowers in early October. (See Figure 5.3 for a photo taken in mid-September.) The flowering *M. ×giganteus* stands out above the *M. sacchariflorus* allowing easy visualization of the first-year clumps. The *M. sacchariflorus* is also senescing. View is from south to north showing the two series of four plots in a checkerboard pattern. Gaps in the *M. ×giganteus* plot are where propagules failed. Photo taken in mid-October.

that year 2 summer drought periods, each of over 1-month duration, may have reduced the second-year yield although the overall annual precipitation was 88 cm. Alternatively, the crop may not have closed canopy to fully exploit resources (Moyer, 2017).

5.3 Site Characterization

For any site it is essential to characterize basic properties. Unless you intend to do research and know some specific details about previous contamination of the site, it is really just a matter of the standard agronomy questions. What are soil pH, texture (percentage sand, silt, clay), organic C, cation exchange capacity, and available nutrients (N, P, K, and micronutrients)? Are there problems of weeds, excess or insufficient water, any hazards from unexploded ordnance? Also very important is whether the site was recently treated with persistent herbicides that could interfere with the establishment of *M. ×giganteus*. If the land is marginal but being used for agriculture, it may be important to know the kind of crop being managed in the years prior to the establishment of Miscanthus planting. Legumes may supply carry-over nitrogen. Long-term grasslands might have relatively few aggressive weeds that are common in highly fertilized crop rotations. Following heavily fertilized crops such as maize, omitting N fertilizer may help reduce weed growth in the first year.

If there is known contamination, you will need analysis of extractable and bioavailable levels of the contaminants. Do some plants grow success-fully on the site now? What type of soil is it – natural, reclaimed after min-ing, a result of extreme grading as for an airfield, or technosol (something manufactured somehow, such as waste dumps, dredged sediments, or mine wastes, combined with other materials)? Is it a uniform area or heteroge-neous, for instance, with many different smaller contaminated areas, as in an abandoned pesticide storage area? Does soil texture changes with depth? Are there known health hazards for workers? Can you have some certainty of access for multiple seasons to establish a harvestable perennial crop?

For a research plot one should do a more detailed investigation in order to later verify changes produced during Miscanthus growth. For instance, to validate claims of successful remediation of organic contaminants it is essen-tial to have good information on the starting concentrations of the contami-nants of concern across the site. This means using a systematic regimen of sampling to appropriate depths and with a spacing suitable to the expected variation or heterogeneity of the site. For a large area uniformly contami-nated, perhaps 10 samples ha^{-1} is sufficient; for a more heterogeneous site in an area where there are many exposed receptors (people or animals) one may need more than 100 samples ha^{-1}. This is especially important if the contami-nants of concern are taken up and stored in Miscanthus because they could spread the exposure beyond the site when harvested and transported, and then processed in some way.

5.4 Plant Nutrition and Supplementation

Soil fertility amendments, if any are needed, will vary with the results of initial characterization. These may include mineral or organic fertilizers, or possibly trace elements such as B or Mo. Organic fertility amendments often are animal or treated human wastes which typically contain trace elements. In the US, "biosolids" is a term used to describe the final product of a residential sewage waste treatment plant. They are closely regulated, much more so than animal wastes. Biosolids are often stabilized with additives such as lime or iron. They are high in organic carbon and the pH is usually near neutral giving efficiency in sorbing metals. Because there is a reasonably high concentration of total N in such waste (4%–8%), biosolids often can serve as an adequate N source for a crop (Evanylo, 2006). Levels of P and K are often beneficial to add to the soil, but the applied N rate is generally used in determining the proper amounts to apply (USEPA, 1994). In addition, high iron in biosolids will increase trace element sorption capacity of the receiving soil. High organic matter will contribute to both trace elements and organic contaminant retention. Further, high organic carbon will promote microbial activities, enhancing organic contaminant degradation, and nutrient cycling. Biosolids applications are also limited by concentrations of specific toxic elements (e.g., As, Pb, Cr, Cd, Zn); however, regulated biosolids (such as Class A and B biosolids in the USA) contain very low levels of trace elements. The field experiment at Ft. Riley, KS, had one treatment of biosolids in its design (Alasmary, 2020).

Other nutrient amendments of soil might include other sources of organic matter such as composted animal confinement wastes. These are a useful source of N, P, K but in varying amounts depending on the animal source and the diet of the animals, which will be very different for ruminants, nonruminant mammals, and poultry. Waste from nonruminants is often very high in P, to the extent that struvite mineral (magnesium ammonium phosphate) can be recovered in large quantities by intentional treatments (Castro-Diaz S., personal communication). In some settings excess P may be the limiting factor for biosolids application. Optimum pH for Miscanthus growth is about 5.5–8. Acidic soils would benefit by application of lime to raise pH into this range. Both calcium and magnesium are essential for crop growth. Based on harvested material in winter, Miscanthus biomass removes relatively little N, P, or K from soil, about 5 kg t^{-1} of N, 0.5 kg t^{-1} of P, and 7 kg t^{-1} of K (Iqbal et al., 2017). Earlier harvest before leaf fall, nutrient translocation, and rainfall leaching of biomass will remove more, often a lot more. As mentioned above, Nagara CV of *M. × giganteus* was noted to retain 3/4 of the leaf N when injured by fall freezing (Sage et al., 2015). Hence an early harvest may remove 20 kg t^{-1}.

Very often it has been observed that added N does not benefit a Miscanthus crop in the planting year. In some situations, added N is beneficial in later years. As recognized by Lee et al. (2017), some locations receive higher levels of N deposition from the atmosphere than others do. Places with high-density animal production may release higher amounts of ammonia to the atmosphere from hydrolysis of urea. At Konza prairie, a relatively "pristine" preserve, near Kansas State University, nitrogen deposition in rainfall and particulates amounted to ~10 kg ha^{-1} $year^{-1}$ over the first 18 years of the 21st century, although there appears to be a downward trend to <9 kg ha^{-1} $year^{-1}$ in the past decade. Lee et al. (2017) suggested that deposition rates were significantly higher in Europe. A reported value of 14 kg ha^{-1} $year^{-1}$ for Germany may be found at Schaap et al. (2017). For the UK, the historic trend is downward in this century and reported as about 10 kg ha^{-1} $year^{-1}$ in 2010 (Tomlinson et al., 2011). For some regions of the UK deposition rates were at least two-fold higher in the 1990s when early studies of Miscanthus were done. The reported deposition rates could support a yield of several Mg ha^{-1} without depleting soil N at all. This is an example of large geographic variability independent of typical climatic rainfall and temperature differences and geographic factors such as latitude and altitude.

Lee et al. (2018) observed quite variable responses across sites and years in the US. Added N at 60 or 120 kg ha^{-1} was rarely detrimental, and sometimes beneficial. The work of Lee et al. (2017) in Illinois (~40 N latitude) showed that very high yields of biomass were obtained with added N, at rates of 60 and 120 kg ha^{-1} year^{-1}, but the higher rate gave no improvement over the 60 kg ha^{-1} rate. Yields were reasonable in the 0 N control plots, but averaged half the yield of the N-fertilized plots over years 2011–2014, where drought may have affected yields (see discussion above under *water*). In other studies, mass balance calculations suggest that the crop may fix atmospheric N by a symbiotic association (Davis et al., 2010). Bioavailability of adequate P may be decreased at high pH in soils with abundant calcium. This combination produces calcium hydroxyapatite, which is quite insoluble at pH 8. Miscanthus extracts K efficiently from soil, but it seems to be a common practice to add a small supplement proportional to amounts removed with the crop, especially after several years of harvest.

A significant concern may be overfertilization of Miscanthus, if it is intended for combustion. This is because high K and high Cl lead to corrosion and slag formation in the combustion chamber. It has been reported that use of large amounts of biosolids can result in plants taking up excess levels of these elements, far above disagree strongly what is essential for growth, lowering the end use value of the biomass (Kołodziej et al., 2016). Those authors applied "sewage sludge" one time in amounts up to 60 dry matter t ha^{-1} and observed that peak productivity was reached at about 10–20 t ha^{-1} in later harvests, but all levels appeared inhibitory in the first year. Nitrogen (total) levels in the biosolids were 7.45% with relatively high ammonium N of 2.35% at the time of application and tilling into 40 cm depth during the autumn prior to planting Miscanthus. The ammonium presumably volatilized or nitrified over winter. If not, it might be toxic at up to 250 mg kg^{-1} of the top 40 cm soil. The cause of first-year inhibition by sewage sludge is unclear, because there were not high concentrations of toxic trace elements or sodium according to data presented by the authors. The Zn level in the sludge was 1000 mg kg^{-1}, which when applied at 60 Mg ha^{-1} would increase the Zn level of soil by only 15 mg kg^{-1} for the top ~40 cm.

A rather different approach was taken by Dubis et al. (2020) who grew Miscanthus in Poland for 5 years with high applications of mineral fertilizer (90 kg ha^{-1} of N, 80 kg ha^{-1} of P_2O_5, 120 kg ha^{-1} of K_2O), and only then, after-stable yields were obtained, compared doses of 100 or 160 kg ha^{-1} of N applied as sewage sludge (~13 and 20 Mg ha^{-1}, varying by year) versus mineral N at comparable rates and 50 kg ha^{-1} of P_2O_5 and 100 kg ha^{-1} of K_2O applied to the mineral fertilizer treatments. Overall, yields varied only ~20% over 6 years, between 17 and 22.5 t ha^{-1}, presumably as a function of weather, with control plots consistently ~20% lower than three treatments which did not differ from one another, and about 10%–15% lower than the lower level of applied sludge. This experiment indicated that for this soil, adding back N at the level that it is withdrawn by harvest is beneficial. Harvest was done relatively early in autumn, when only 1/3 of leaves up the stem had dried, for use in silage for biogas production. Thus, the offtake of N was much greater than it would be in a late winter harvest. For instance, Kołodziej et al. (2016) saw a decrease of nearly two-fold in ash content of *M. ×giganteus* when comparing autumn with spring harvested material. Iqbal et al. (2017) showed that offtake varies with harvest date at multiple locations with multiple CVs, consistent with the estimate that early harvest before nutrient translocation would remove 100–160 kg year^{-1} of N, while later winter harvest would decrease this to 1/2 or 1/3 the amount. For combustion, lowered total mineral content is very important, whereas for biogas production it is not.

5.5 Role of Soil Amendments

Inorganic commercial fertilizers contain N, P, K, as their main nutrients in the form of various salts, sometimes with Ca, Mg, Zn, S, B, or trace elements added for specific soil types. These along with different organic fertilizers (compost, ash, manure, activated carbon) have historically been the primary soil amendments (Antonkiewicz et al., 2019; Boakye-Boaten et al., 2016; Lehmann et al., 2003). In the last 25 years different wastes like sewage sludge and digestate are becoming popular as soil amendments as well (Antonkiewicz et al., 2020; Kirchmann et al., 2017; Tabak et al., 2020). In addition, biochar, the solid material obtained from the carbonization of biomass/waste or through pyrolysis, is currently proposed as both a soil amendment and carbon sequestration medium (Agegnehu et al., 2016; Faria et al., 2018; Lehmann et al., 2006). Application of soil amendments boosts the soil fertility balance and improves soil quality, resulting in increased crop yields (Hu et al., 2018; Humentik et al., 2018). Improvements in soil fertility result in greater uptake of macronutrients and micronutrients by plants, mainly in the second cropping season, and higher biomass productivity. Soil benefits include optimizing soil pH, increasing moisture holding capacity, attracting more beneficial fungi and microbes, improving cation exchange capacity, and retaining nutrients. These benefits have been shown to increase yield in biomass and crops under variable conditions (Chan et al., 2008). One obvious energy cycle is to grow Miscanthus, pyrolyze it to recover energy, and use the residual biochar as a source of minerals and carbon for Miscanthus cultivation again.

The impacts of application of different soil amendments to the production of energy crops on regular agricultural soils are well represented in the literature. However, the improvements of biomass production when these crops are produced on marginal or contaminated soils are not researched so well. Only a few publications have evaluated changes in the phytoremediation parameters of the second-generation biomass crop Miscanthus when it is growing on contaminated soil receiving different soil amendments (Alasmary, 2020; Ameen et al., 2018; Kucharski et al., 2005; Mamirova et al., 2020). See Chapter 3 for some specific examples.

5.5.1 Impact of Soil Amendments on the Phytoremediation of Soil Contaminated by Organic Substances

Low molecular weight compounds. Plant roots secrete a wide range of chemical compounds: multicarboxylic organic acids including aconitic, citric, malic, malonic, oxalic, succinic, and tartaric acids; sugars and sugar conjugates; amino acids and peptides; phenolics, some of which are allelopathic; and diverse enzymes. Often the exudation results from complex interactions with the root microbiome (Korenblum et al., 2020). Exudation is sometimes initiated by lack of nutrients, pollutant toxicity, or anoxia (Dakora & Phillips, 2002; Zeng et al., 2008). Root exudates thus serve to interact directly with contaminants, or indirectly by their influence on the microbiome. See Chapter 2 for more details.

In 1995, Hülster & Marschner proposed a hypothesis that root exudates can bind with persistent organic pollutants in soil and form a more hydrophilic complex which can be more easily absorbed by roots and translocated to aboveground biomass. Campanella & Paul (2000) supported this hypothesis, finding that *Cucurbita pepo* and melon (*Cucumis melo*) root exudates bind dioxins and furan molecules facilitating their translocation to aboveground biomass. They suggested that at least some part of this mix of carrier molecules was proteinaceous. The impact of organic acids, citrate, and EDTA (ethylenediaminetetraacetic acid) on the p,p'-DDE (p,p'-dichlorodiphenyl-1,1-dichloroethene) uptake by *Cucurbita pepo, Trifolium incarnatum, Brassica juncea, Vicia villosa,* and *Lolium multiflorum* was investigated by White et al. (2003) and White and Kottler (2002). They observed significant increases in uptake of p,p'-DDE for *Cucurbita pepo* (succinic acid – 19%; tartaric acid – 27%; malic acid – 31%; malonic acid – 36%; oxalic acid – 45%; citric acid – 58%; EDTA – 80%) and for *Trifolium incarnatum, Brassica juncea,* and *Vicia villosa* (citrate – 39%). Citrate also chelated metals, altering their bioavailability. More recently a 17 kDa protein of the major latex protein class was identified in the xylem sap of *C. pepo* and shown by genetic means to be correlated with enhanced translocation of broad classes of POPs (Inui et al., 2013). So, it may be that there are multiple facilitators of their uptake at multiple steps.

Surface-active compounds. Surfactants are chemical compounds that decrease surface tension. Surfactants can reduce the hydrophobicity of organic compounds, for example, nonionic surfactants decreased the hydrophobicity of polychlorinated biphenyls in a soil-water system (Park & Boyd, 1999). There are surfactants of chemical (Tweens, Polysorbate, Triton) and biological (e.g., rhamnolipids) origins. Gonzalez et al. (2010) reported that adding Tween 80 (nonionic) to contaminated soils effectively enhanced p,p-DDT, p,p-DDE, and α-cypermethrin solubility while adding sodium dodecyl sulfate (anionic) increased the solubility of two other pesticides (α-endosulfan and endosulfan sulfate).

Rhamnolipids, also called biosurfactants, are glycolipids produced by *Pseudomonas, Burkholderia*, and other genera (Abdel-Mawgoud et al., 2010). Amendment of soil contaminated with p,p'-DDE by adding biosurfactants increased accumulation of pesticides in roots, leaves, and fruits of *Cucurbita pepo* ssp. *Pepo* (hyperaccumulator) and *C. pepo* ssp. *Ovifera* (nonaccumulator) mainly by reducing its net hydrophobicity (White et al., 2006).

Carbon-rich materials. Application of carbon-rich materials (biochar, activated carbon, lignite, etc.) in a phytoremediation process aims to stabilize organic pollutants by reducing their bioavailability (Denyes et al., 2012). A comparative study on potential of biochar and activated carbon to decrease the bioavailability of polychlorinated dibenzo-*p*-dioxins and -furans showed reduction of their bio-uptake in earthworms by 51.6%–90.3% (Chai et al., 2012). In the Chai et al. (2012) research, contaminant reduction was higher in soil treated by activated carbon whereas in Denyes et al. (2013) it was almost the same. The effectiveness of carbon-rich materials also depends on the way they were added to the system. In a mechanically mixed system (24 hours, 30 rpm rotation in a drum) activated carbon reduced polychlorinated biphenyl levels 1.7 and 1.4 times more efficiently in the plant *Cucurbita pepo* and earthworm *Eisenia fetida*, respectively, while the effectiveness of biochar was higher by 2.0 and 1.7 times, respectively, as compared to simply digging the material into soil at a contaminated site (Denyes et al., 2013). Contact of contaminant and sorbent is slow in natural systems.

Nanoparticles. Nanoremediation is a relatively new area of environmental biotechnology, based on the ability of Ag, Au, Mg, and Fe nanoparticles to facilitate dehalogenation of halocarbon pesticides. Nanoparticles can either directly react with contaminant or participate in its conversion into less toxic forms (Adeleye et al., 2013). Nanoparticles have been shown to be an efficient amendment able to degrade 100% of DDT in various matrices (Tian et al., 2009). Applying zero-valent iron nanoparticles (nZVIs) for DDT dechlorination in water and soil systems showed that the potential of nZVIs to decompose DDT was higher (92%) in water than in soil (22.4%) over the same time (El-Temsah et al., 2016). Unfortunately, different modes of preparation of the nZVI yield different eco-toxicity, which is nontrivial when it is applied in large amounts to obtain effective degradation of POPs such as DDT. The nZVIs were able to degrade lindane (γ-HCH (hexachlorohexane)) within 24 hours to γ-3,4,5,6-tetrachlorocyclohexane (an unstable intermediate) (Elliott et al., 2009). The high rates were largely driven by the very high surface area of nanoparticles, compared to larger ZVI particles.

Initial research combining nano- and phytotechnology to restore soil polluted by a chlorinated pesticide (endosulfan) included three tropical plant species: *Alpinia calcarata*, a large monocot in the ginger family; *Ocimum sanctum*, a perennial dicot type of basil; and *Cymbopogon citratus*, a true grass commonly called lemon grass, grown without or with nZVIs in pots (Pillai & Kottekottil, 2016). These three species had been selected from 11 species grown at a level of 1 mg kg^{-1} endosulfan. The endosulfan concentration in soil for testing nZVI effects contained 1140 μg kg^{-1}, while the optimized level of nZVI was 1 g kg^{-1}. Ten mL kg^{-1} of Tween 80 was used to disperse the nZVI; there were three replicates of six treatments. *A. calcarata* showed better phytoremediation potential in comparison to *C. citratus* and *O. sanctum* in both treatments: on the 7th day it removed 52% of endosulfan from pots without and 82% for pots with nZVIs; while on the 28th (last) day, endosulfan removal was 81% without and 100% with nZVIs at a level of 1 g kg^{-1}. In the case of *O. sanctum*, addition of nZVIs to the system led to endosulfan removal increasing 3.7-fold over the plant alone (72% vs 21%) on day 28. With *C. citratus*, the main nZVI enhancement of endosulfan removal was observed after seven days (63% vs 5%). By day 14 the difference was 81% vs 60% and on day 28 with nZVIs removal was 86% and without nZVIs it was 65% (Pillai & Kottekottil, 2016). The patterns of rate difference are very nonlinear indicative of a perhaps second-order rate. Thus, the combined technology of nano- and phytoremediation is one of the promising areas for the remediation of organochlorine pesticides.

5.5.2 Impact of Soil Amendments on Miscanthus Production in Postmilitary Soil

The impact of soil amendments on Miscanthus biomass parameters was evaluated under field conditions with the crop cultivated in *postmilitary soil, slightly contaminated by trace elements* in Dolyna, Western Ukraine. Three different soil amendments were tested: lime GOST 14050-93; mineral fertilizer "Smolokot" (the nutrient content is N:P:K = 8:8:12, +7% MgO + microelements); organic fertilizer "Agrolife" (N:P:K = 10:10:10 with chicken compost); and their mixture. Different fertilizers were mixed with the soil in a dose of 40 g per plant, while lime was 30 g per plant. The presoaked *M. × giganteus* rhizomes were planted directly into this mixture at a depth of 10 cm. Results are presented in Figures 5.5–5.7. Treatment A1 is a control, with rhizomes soaked in water, planted without amendments to soil. A2–A5 all have rhizomes soaked in Charkor plant growth stimulant added 4 ml L^{-1} water. A2 has lime, A3 has AgroLife fertilizer, A4 has Smolokot fertilizer, and A5 has all three. All plots were 25 m^2, with A1, A2, A5 receiving 56 propagules, with A3, and A4 having 42. There were three replicates of each treatment.

Through analysis of results from 3 years monitoring of *M. × giganteus* development in the contaminated military soil with application of different soil amendments, it may be concluded that the greatest effect was obtained when all three amendments were applied and the biggest influence was observed for dry mass production (Figure 5.7). The impact was small on crop height (Figure 5.6).

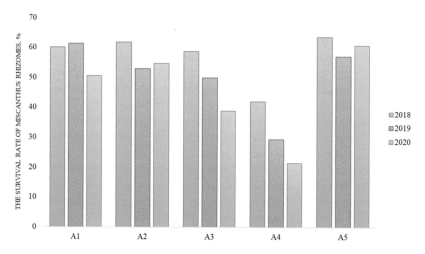

FIGURE 5.5
The mean winter survival rate of *M. ×giganteus* grown in the "postmilitary" soil with amendments, measured in spring of indicated year.

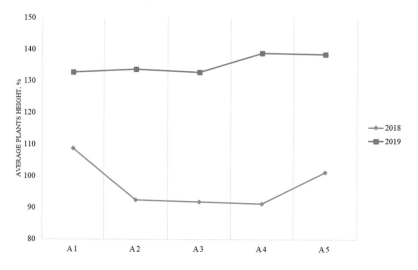

FIGURE 5.6
Plant heights of *M. ×giganteus* when grown in "postmilitary" soil with amendments, measured in autumn of indicated year.

5.5.3 Impact of Soil Amendments on Miscanthus Biomass Production in Contaminated Postmining Soil

M. ×giganteus was tested for revitalization of postmining land with biomass production (Kharytonov et al., 2019). The research soil consisted of the mixture of loess-like loam and red-brown clay which had passed through a long-term phytomelioration stage. The soil humus content was about 1.5%, and the

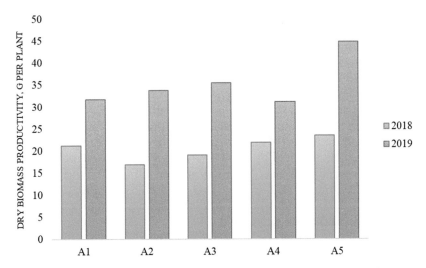

FIGURE 5.7
Mean dry biomass per surviving plant of *M. ×giganteus* for different amendments in the postmilitary soil.

ratio of humic and fulvic acids was 0.2–0.5, which indicates a weak humus accumulation and active destruction of the soil mineral part. *M. ×giganteus* showed sufficient tolerance and good enough growth and development in this postmining soil during 2 years of experimentation (Figure 5.8).

In order to determine the impact of amendments on the growth and development parameters of *M. ×giganteus*, different amendments were applied: mineral fertilizer with a balance of nutrients N_{60}:P_{60}:K_{60} kg ha^{-1}; ash of sunflower hulls or sewage sludge each in the amount 10 t ha^{-1}; a mixture of ash and sewage sludge (total 10 t ha^{-1}); a double dose of sludge (20 t ha^{-1}). The amendments were incorporated into soil in a dry form once before planting.

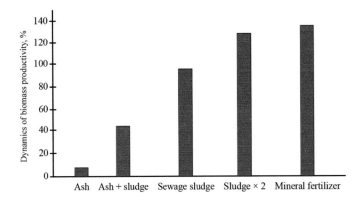

FIGURE 5.8
Miscanthus biomass productivity (% increase over control plots) in the postmining soil with different amendments. (Modified from Kharytonov et al., 2019.)

The results showed that the application of amendments positively affected the growth parameters of Miscanthus (Figures 5.8). The biggest effect was observed when plants grew in soil amended by mineral fertilizers, while the application of ash showed minimal increases of the growth parameters.

The increase of growth parameters promoted the enhancement of above ground biomass yield, and the degree of this enhancement was correlated with growth parameters, though for each the enhancement was different. In the case of ash application, yield increased by only 6.5% compared with the control, while the addition of mineral fertilizer or sewage sludge significantly increased the biomass yield by 2–2.3 times. As a result, Miscanthus productivity in the second year was 11.6 t dry matter (DM) ha^{-1} with sludge and 11.9 t DM ha^{-1} with mineral fertilizer. This research confirms earlier reported data (An et al., 2011; Kołodziej et al., 2016) that sewage sludge is conducive to increasing Miscanthus biomass yield.

The results illustrate that application of soil amendments can increase plant productivity by 50%–140% and that one can obtain Miscanthus yields similar to arable lands. Special attention should be paid to using sewage sludge as a promising substitute for organic fertilizers while growing *M. × giganteus* on postmining lands.

5.6 Geography and Soil Types

In the extensive review of Heaton et al. (2010) there is a very useful discussion of the impact of geology and soil type on *M. × giganteus* productivity. They cite several studies which indicate that an optimum soil is neither too sandy nor too dense, but intermediate in texture. This increases both water-holding capacity and aeration. Depth of soil is also important as water capacity depends on both texture and depth of soil. Capacity may be underestimated if a site is underlain by very porous rock which can effectively release water to the overlying soil during drought events or which can be reached by plant roots. Reasonable levels of clay improve capacity to store essential minerals including K, Ca, Mg, and cationic trace elements. Lee et al. (2017) attributed the N response observed in their long-term study in part to the sandy nature of the soil being used and its relatively low retention of N. Organic matter increases the formation of aggregates and improves the overall tilth of the soil. In its native habitat of Japan, *M. sinensis* is able to pioneer and grow on volcanic ash-derived soils, and tolerates high levels of Al at low pH (Stewart et al., 2009). It contributes large amounts of organic matter to the developing soil in those conditions, where frequent fires over millennia have also contributed biochar.

As an example of successful production of Miscanthus on a potentially unfavorable site, we may consider the work of Skousen and co-workers in West Virginia (Scagline et al., 2015; Skousen et al., 2013). The site (38°49′ N,

80°11′ W) had been surface-mined, restored with mixed overburden, and 15 cm soil covering, fertilized, limed, and planted with mixed grass and legumes 25 years earlier. It had been managed as mown forbs during that time. Five replicate plots of 0.4 ha were planted to each of two types of *M. × giganteus* – public and private from Mendel Biotechnology (unnamed but likely Illinois and Nagara CVs). All plots were planted into a sod previously killed with glyphosate herbicide applied twice, fall and spring prior to planting. Planting density was 12,300 plugs ha^{-1}. Even with a >20% loss of plants this would still produce an appropriate plant density. Other plots received two types of switchgrass, drilled into the killed sod. Yields of switchgrass were 7.9 and 7.3 t ha^{-1} while the private *M. × giganteus* yielded 13.7 and the public 14.4 t ha^{-1} in year 5. The private CV increased yield much more rapidly over years than the public, reaching 22 t ha^{-1} in year 3. However, all measures of Miscanthus had large variation (SD ~50%, N = 5 plots) likely because all yield measures were based on random selections of six plants, from large (0.4 ha) plots.

These *M. × giganteus* yields are very like those reported in the south-central Virginia Piedmont region (~36°56′ N, 79°24′ W) in the same years by Lee et al. (2018) and Battaglia et al. (2019). Switchgrass yields in West Virginia (WV) were better than unfertilized soil of the Virginia (VA) portion of Lee et al. (2018) studies, but lower than on fertilized plots. For switchgrass, six 0.21 m^2 quadrats were sampled on each of the WV switchgrass plots, but again the SD was very large. A critical methodological difference is that the studies of *M. × giganteus* coordinated by Lee et al. (2017) used a 4 m^2 sample from the center of a plot of 10 × 10 m. With N = 4 for each different nitrogen fertilization level, the reported Standard Error of the Mean was much smaller, closer to 10%. In the VA switchgrass study, there was a positive, usually large, response to added N each year. Their yields were measured by complete biomass harvest on full plots of at least 0.5 ha.

A very useful European example is the work of Jeżowski et al. (2017) who grew *M. × giganteus* on lignite mine spoils, with sewage sludge and mineral fertilizer supplements. Three years prior to planting Miscanthus, a mix of Medicago species was planted on the site which was roughly leveled. The wet sludge (1 Mg per 5 × 5 m treatment plot) was applied in the autumn prior to the Miscanthus planting and was incorporated into the soil to a depth of ~30 cm. This corresponds to 400 t ha^{-1} wet wt, 80 t dry matter wt (DM). Rhizomes were planted on 1 m centers, 25 per plot, in a randomized, three complete blocks design. Treatments were D_0 = no addition, D_1 = addition of ~80 Mg ha^{-1} (DM) of municipal sewage sludge, D_2 = sludge + 200 kg ha^{-1} commercial fertilizer, D_3 = sludge + 400 kg ha^{-1} of commercial fertilizer. With a composition listed as 13:19:16 for N:P:K, the added fertilizer makes a modest N contribution of ~26 kg ha^{-1} to D_2 and ~52 kg ha^{-1} to D_3. Total N for the sludge was 43.4 kg Mg^{-1} DM (4.3%) for an applied rate of over 3400 kg ha^{-1}, although most would be organic N. If this mineralizes at a rate of 2% per year it contributes all the N needed by the amount of biomass produced, over a long time.

Biomass yields increased year by year, but D_1–D_3 did not differ significantly from one another in any year. Within treatments yield estimates varied ±20%. "From the center of each plot, six randomly selected plants were collected". Yields in year 3 were ~9 Mg ha^{-1} for the control D_0 and ~15 Mg ha^{-1} for the three treatments with sludge. This represents a positive use for a marginal land. The authors estimated that it might take 7–10 years to become profitable as a crop, because of establishment and harvest costs.

Overall, there is good evidence that geographic location is very important to potential *M. ×giganteus* yields, as reviewed by Heaton et al. (2010), independent of fertility of soil, addition of fertilizers, and other nutrient amendments. Latitude influences time of flowering and yield. Higher latitudes delay flowering. Rainfall patterns (timing and amounts) and temperature regimes determine the regions where rain-fed crops can be successfully grown for maximum production. Continental vs oceanic climates markedly affect stability of yields. For the U.S. there are adaptability maps showing likely zones of relative yield, based on models of *M. ×giganteus* growth patterns. There have been enough regional studies to indicate that in some locations alternatives other than *M. ×giganteus* are more productive at least for a few years of study (Smith et al., 2015b). There have not been sufficient long-term studies to say how they would fare over the course of decades. Similar maps have also been developed with alternate bioenergy crops including maize, switchgrass, sorghum, and select sugarcane CVs identified as "Energy Cane" (Matsuoka et al., 2014). For total biomass production in the south of US, *M. ×giganteus* cannot compete with Energy Cane.

Very sophisticated models of growth patterns for *M. ×giganteus* in Europe have been developed. Early examples are discussed in Heaton et al. (2010). These have good predictive values when key features including latitude, hours of light, water, and temperature are input to the models.

5.7 Role of Plant Growth Regulators in Production of *M. × giganteus*

Plant growth regulators (PGRs, earlier term "phytohormones") are a group of treatment substances used for enhancing plant growth (Procházka & Šebánek, 1997). There are five main classical groups of natural PGRs: auxins, cytokinins, gibberellins, ethylene, abscisic acid; in addition, there is a class of "new plant hormones" formed by brassinosteroids, salicylic acid, jasmonates, and strigolactones. For each class there are substances chemically or biologically synthesized and used as mimics or inhibitors (agonists and antagonists in biochemical terminology). Also, oligosaccharides, systemin, polyamines, reactive oxygen species, and reactive nitrogen species including nitric oxide possess activities similar to those of plant hormones in various systems (Bhattacharyya & Jha, 2012; Chen et al., 2009; George et al., 2008; Ponomarenko et al., 2010, 2017).

Applications of PGRs have been studied in Europe for increasing the production of different agricultural crops: wheat, maize, sunflowers (Ponomarenko et al., 2010; Tsygankova et al., 2013a). A total of 14 different PGRs produced by Agrobiotech Company (Ukraine), based on natural active ingredients, were tested on *Sorghum bicolor* L., sunflowers, wheat, and maize production, and the best results were received for two PGRs, Stimpo and Regoplant. While growing sunflowers, treatment with these PGRs increased the plant weight by 6.4% and the number of seeds in a head by 13.0%. In the case of maize production such treatment increased plant height by 11.5%, leaf surface area by 12.7%, and root length by 12.5%.

The application of PGRs to plants that grow on contaminated land may protect the photosynthetic apparatus from oxidative shock induced by contaminants, increase root length, increase shoot growth, enhance the transpiration rate and yield (Israr et al., 2011; Liphadzi et al., 2006). The use of PGRs usually results in benefit with little risk of negative environmental effects, and PGRs may also boost plant immunity to pests and pathogens (Ponomarenko et al., 2013; Tsygankova et al., 2013b). Only a few publications exist on the impact of PGRs in *M. × giganteus* production, when a crop was grown on regular agricultural soil during one vegetation season (Zinchenko, 2013). Treatment of *M. × giganteus* before planting with three PGRs Regoplant, Emistin, and Agrostimylin produced by AgroBiotech Company (Ukraine) increased the activity of photosynthesis and improved the survival rate. The treatment by PGRs stimulated plant development by increasing the number of stems and the overall crop height. The application of the same PGRs, Regoplant, Emistin, and Agrostimylin, increased the photosynthesis rate when *M. × giganteus* was cultivated in the soil contaminated by radionuclides and gave an increase in the biomass at harvest of about 18.8%–25.3% (Zinchenko et al., 2016; Zinchenko et al., 2009).

5.7.1 Lab Research on Impact of PGRs on Phytoremediation with Biomass Production Using Soils from Military Sites Contaminated with Trace Elements

The impact of two PGRs, Stimpo and Regoplant, was tested in pots, with four soils taken from two military sites of different origins (Dolyna, Ukraine and Mimon, Czech Republic, two from each area). Soils in both cases were slightly contaminated by heavy metals (Nebeská et al., 2019). PGRs were produced by AgroBiotech Company (Ukraine) and consisted of a balanced composition of biologically active compounds, namely, analogues of phytohormones, amino acids, fatty acids, oligosaccharides, microelements, and bioprotective compounds. The treatment of rhizomes was done before planting and by spraying on the biomass during growth. The results obtained (Nebeská et al., 2019) showed that the main factor driving the increase of biomass parameters was the agricultural characteristics of the soil: the better it was the greater the effect of PGRs treatment was (case of Dolyna soil). Conversely, the effect was almost negligible in soils poor in nutrients and organic matter (case of Mimon soil). Between the two tested PGRs the effect of Regoplant was more obvious. The best results were obtained with combined treatment of application to rhizomes before planting and spraying of biomass during the vegetation season (Nebeská et al., 2019).

The impact of PGR treatment on phytoremediation parameters was evaluated as well. When *M. ×giganteus* was growing in soil richer in nutrients the process of metals uptake was in accordance with the general trend for Miscanthus (Pidlisnyuk et al., 2014) and recognized as phytostabilization. That is, the majority of the metals accumulated in the roots. The bioaccumulation behavior of the monitored metals was different when a crop was grown in sandy soil with poor nutrient content. While the nonessential elements Cr and Pb were as expected dominantly accumulated in roots, Ni was not detected and biogenic elements (Mn, Cu, Zn) were more intensively taken up into shoots in comparison with roots. This was attributed to stress caused by deficient soil characteristics (Kabata-Pendias & Pendias, 2001; Nebeská et al., 2019).

5.7.2 Field Research on Impact of PGRs on Biomass Parameters of *M. × giganteus* during Field Production on the Marginal and Slightly Contaminated Lands

Two PGRs, Stimpo and Charkor, were evaluated in the field conditions and tested for production of Miscanthus on marginal agricultural land in the Central Ukraine (Pidlisnyuk & Stefanovska, 2018). The dose for the treatment was the same as in the Lab experiment (Nebeská et al., 2019). The results received after first year of plant's growth showed about 20% increase of biomass when rhizomes were treated by Stimpo, and about 28% increase when rhizomes were treated by Charkor (Table 5.1).

TABLE 5.1

Bioparameters of *M. × giganteus* When a Crop Was Produced on the Marginal Agricultural Land with Treatment of Rhizomes by PGRs

Treatment	Structure of Harvest (%)		Harvest, Dry Biomass/Plant (g)	Increase (%)
Water (control)	Leaves	21.0	6.5	-
	Stems	79.0		
Stimpo plantation	Leaves	21.0	7.8	20
	Stems	79.0		
Charkor 1 plantation	Leaves	26.0	8.3	28
	Stems	74.0		
Charkor 2 plantation	Leaves	26.0	8.3	28
	Stems	74.0		

Source: Modified from Pidlisnyuk and Stefanovska (2018).

In another experiment in Dolyna, Western Ukraine, the PGR Charkor showed the best results in comparison with Stimpo and Regoplant during multiple years of *M. ×giganteus* production on the military soil slightly contaminated with trace elements (Figures 5.9–5.11). Application of Charkor improved the overwinter survival rate, average plant height at harvest, and dry biomass productivity in comparison with control treatment by water or two other tested PGRs: Stimpo and Regoplant.

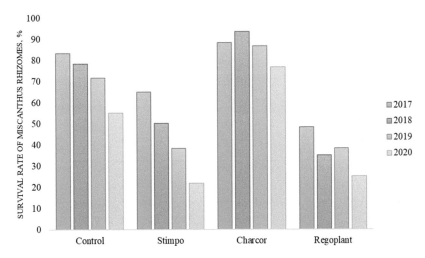

FIGURE 5.9
Winter survival rate of *M. ×giganteus* plants, until spring of the indicated year.

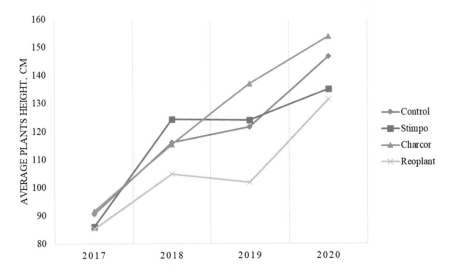

FIGURE 5.10
Average plant height of *M. ×giganteus* at harvest, cm.

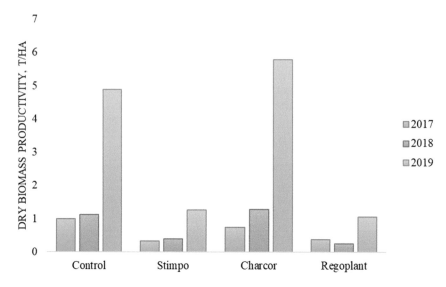

FIGURE 5.11
Dry biomass productivity of *M.* × *giganteus,* Mg ha⁻¹.

PGRs can be recommended as substances which improve the biomass parameters of *M.* × *giganteus* when the crop is growing on marginal or slightly contaminated soils of military origin. The best results were received for Charkor which showed the biggest increase of biomass along with improvement of the survival rate compared to other tested PGRs, Stimpo and Regoplant. When PGRs were applied to soil depleted of nutrients the effect of these substances was very little.

References

Abdel-Mawgoud, A. M., Lépine, F., & Déziel, E. (2010). Rhamnolipids: Diversity of structures, microbial origins and roles. *Applied Microbiology and Biotechnology,* *86*(5), 1323–1336. https://doi.org/10.1007/s00253-010-2498-2.

Adeleye, A. S., Keller, A. A., Miller, R. J., & Lenihan, H. S. (2013). Persistence of commercial nanoscaled zero-valent iron (nZVI) and by-products. *Journal of Nanoparticle Research,* *15*(1), 1418. https://doi.org/10.1007/s11051-013-1418-7.

Agegnehu, G., Bass, A. M., Nelson, P. N., & Bird, M. I. (2016). Benefits of biochar, compost and biochar–compost for soil quality, maize yield and greenhouse gas emissions in a tropical agricultural soil. *Science of the Total Environment,* *543*, 295–306. https://doi.org/10.1016/j.scitotenv.2015.11.054.

Alasmary, Z. (2020). *Laboratory- to field-scale investigations to evaluate phosphate amendments and Miscanthus for phytostabilization of lead-contaminated military sites* [PhD Dissertation, Kansas State University]. https://krex.k-state.edu/dspace/handle/2097/40676.

Ameen, A., Tang, C., Han, L., & Xie, G. H. (2018). Short-term response of switchgrass to nitrogen, phosphorus, and potassium on semiarid sandy wasteland managed for biofuel feedstock. *BioEnergy Research, 11*(1), 228–238. https://doi.org/10.1007/s12155-018-9894-3.

An, G. H., Lee, S., Koo, B. C., Choi, Y. H., Moon, Y. H., Cha, Y. L., Bark, S. T., Kim, J. K., Kim, B. C., & Kim, S. P. (2011). Effects of application of solidified sewage sludge on the growth of bioenergy crops in reclaimed land. *Korean Journal of Crop Science/Hanguk Jakmul Hakhoe Chi, 56*(4), 299–307. https://doi.org/10.7740/kjcs. 2011.56.4.299.

Anderson, E. K. (2010). *Herbicide phytotoxicity response and eradication studies on Miscanthus × giganteus* [MS Thesis, University of Illinois]. http://hdl.handle.net/2142/18498.

Anderson, E. K., Arundale, R., Maughan, M., Oladeinde, A., Wycislo, A., & Voigt, T. (2011). Growth and agronomy of *Miscanthus × giganteus* for biomass production. *Biofuels, 2*(1), 71–87. https://doi.org/10.4155/bfs.10.80.

Anderson, E. K., Lee, D., Allen, D. J., & Voigt, T. B. (2015). Agronomic factors in the establishment of tetraploid seeded *Miscanthus × giganteus*. *GCB Bioenergy, 7*(5), 1075–1083. https://doi.org/10.1111/gcbb.12192.

Antonkiewicz, J., Kołodziej, B., Bielińska, E. J., & Popławska, A. (2019). The possibility of using sewage sludge for energy crop cultivation exemplified by reed canary grass and giant miscanthus. *Soil Science Annual, 70*(1), 21–33. https://doi.org/10.2478/ssa-2019-0003.

Antonkiewicz, J., Popławska, A., Kołodziej, B., Ciarkowska, K., Gambuś, F., Bryk, M., & Babula, J. (2020). Application of ash and municipal sewage sludge as macronutrient sources in sustainable plant biomass production. *Journal of Environmental Management, 264*, 110450. https://doi.org/10.1016/j.jenvman.2020.110450.

Aurangzaib, M. (2012). *Performance evaluation of nine varieties of Miscanthus in Iowa* [Master Thesis, Iowa State University]. https://lib.dr.iastate.edu/etd/12690.

Awty-Carroll, D., Hauck, B., Clifton-Brown, J., & Robson, P. (2020). Allelopathic and intraspecific growth competition effects on establishment of direct sown Miscanthus. *GCB Bioenergy, 12*(6), 396–409. https://doi.org/10.1111/gcbb.12680.

Battaglia, M., Fike, J., Fike, W., Sadeghpour, A., & Diatta, A. (2019). *Miscanthus × giganteus* biomass yield and quality in the Virginia Piedmont. *Grassland Science, 65*(4), 233–240. https://doi.org/10.1111/grs.12237.

Bhattacharyya, P. N., & Jha, D. K. (2012). Plant growth-promoting rhizobacteria (PGPR): Emergence in agriculture. *World Journal of Microbiology and Biotechnology, 28*(4), 1327–1350. https://doi.org/10.1007/s11274-011-0979-9.

Boakye-Boaten, N. A., Xiu, S., Shahbazi, A., Wang, L., Li, R., Mims, M., & Schimmel, K. (2016). Effects of fertilizer application and dry/wet processing of *Miscanthus × giganteus* on bioethanol production. *Bioresource Technology, 204*, 98–105. https://doi.org/10.1016/j.biortech.2015.12.070.

Boersma, N. N., Dohleman, F. G., Miguez, F. E., & Heaton, E. A. (2015). Autumnal leaf senescence in *Miscanthus × giganteus* and leaf [N] differ by stand age. *Journal of Experimental Botany, 66*(14), 4395–4401. https://doi.org/10.1093/jxb/ery129.

Boersma, N. N., & Heaton, E. A. (2012). Effects of temperature, illumination and node position on stem propagation of *Miscanthus × giganteus*. *GCB Bioenergy, 4*(6), 680–687. https://doi.org/10.1111/j.1757-1707.2011.01148.x.

Boersma, N. N., & Heaton, E. A. (2014a). Does propagation method affect yield and survival? The potential of *Miscanthus × giganteus* in Iowa, USA. *Industrial Crops and Products, 57*, 43–51. https://doi.org/10.1016/j.indcrop.2014.01.058.

Boersma, N. N., & Heaton, E. A. (2014b). Propagation method affects *Miscanthus × giganteus* developmental morphology. *Industrial Crops and Products, 57,* 59–68. https://doi.org/10.1016/j.indcrop.2014.01.059.

Campanella, B., & Paul, R. (2000). Presence, in the Rhizosphere and leaf extracts of Zucchini (*Cucurbita pepo* L.) and melon (*Cucumis melo* L.), of molecules capable of increasing the apparent aqueous solubility of hydrophobic pollutants. *International Journal of Phytoremediation, 2*(2), 145–158. https://doi.org/10.1080/15226510008500036.

Caslin, B., Finnan, J., & Easson, L. (2011). *Miscanthus best practices guidelines.* http://greenbiomassenergy.com/wp-content/uploads/2020/02/Miscanthus Best Practice Manual 190913.pdf.

Chai, Y., Currie, R. J., Davis, J. W., Wilken, M., Martin, G. D., Fishman, V. N., & Ghosh, U. (2012). Effectiveness of activated carbon and biochar in reducing the availability of polychlorinated dibenzo-p-dioxins/dibenzofurans in soils. *Environmental Science & Technology, 46*(2), 1035–1043. https://doi.org/10.1021/es2029697.

Chan, K. Y., Zwieten, L. V., Meszaros, I., Downie, A., & Joseph, S. (2008). Agronomic values of greenwaste biochar as a soil amendment. *Soil Research, 45*(8), 629–634. https://doi.org/10.1071/SR07109.

Chen, C., Zou, J., Zhang, S., Zaitlin, D., & Zhu, L. (2009). Strigolactones are a new-defined class of plant hormones which inhibit shoot branching and mediate the interaction of plant-AM fungi and plant-parasitic weeds. *Science in China Series C: Life Sciences, 52*(8), 693–700. https://doi.org/10.1007/s11427-009-0104-6.

Clifton-Brown, J. C., Hastings, A., Mos, M., et al. (2017). Progress in upscaling Miscanthus biomass production for the European bio-economy with seed-based hybrids. *GCB Bioenergy, 9*(1), 6–17. https://doi.org/10.1111/gcbb.12357.

Clifton-Brown, J. C., & Lewandowski, I. (2000). Water use efficiency and biomass partitioning of three different Miscanthus genotypes with limited and unlimited water supply. *Annals of Botany, 86*(1), 191–200. https://doi.org/10.1006/anbo.2000.1183.

Clifton-Brown, J. C., Schwarz, K.-U., Awty-Carroll, D., et al. (2019). Breeding strategies to improve Miscanthus as a sustainable source of biomass for bioenergy and biorenewable products. *Agronomy, 9*(11), 673. https://doi.org/10.3390/agronomy9110673.

Cosentino, S. L., Patanè, C., Sanzone, E., Copani, V., & Foti, S. (2007). Effects of soil water content and nitrogen supply on the productivity of *Miscanthus × giganteus* Greef et Deu. in a Mediterranean environment. *Industrial Crops and Products, 25*(1), 75–88. https://doi.org/10.1016/j.indcrop.2006.07.006.

Dakora, F. D., & Phillips, D. A. (2002). Root exudates as mediators of mineral acquisition in low-nutrient environments. In J. J. Adu-Gyamfi (Ed.), *Food Security in Nutrient-Stressed Environments: Exploiting Plants' Genetic Capabilities* (pp. 201–213). Springer, Netherlands. https://doi.org/10.1007/978-94-017-1570-6_23.

Davis, S. C., Parton, W. J., Dohleman, F. G., Smith, C. M., Grosso, S. D., Kent, A. D., & DeLucia, E. H. (2010). Comparative biogeochemical cycles of bioenergy crops reveal nitrogen-fixation and low greenhouse gas emissions in a *Miscanthus × giganteus* agro-ecosystem. *Ecosystems, 13*(1), 144–156. https://doi.org/10.1007/s10021-009-9306-9.

Denyes, M. J., Langlois, V. S., Rutter, A., & Zeeb, B. A. (2012). The use of biochar to reduce soil PCB bioavailability to *Cucurbita pepo* and *Eisenia foetida. Science of the Total Environment, 437,* 76–82. https://doi.org/10.1016/j.scitotenv.2012.07.081.

Denyes, M. J., Rutter, A., & Zeeb, B. A. (2013). In situ application of activated carbon and biochar to PCB-contaminated soil and the effects of mixing regime. *Environmental Pollution, 182,* 201–208. https://doi.org/10.1016/j.envpol.2013.07.016.

Dong, H., Green, S. V., Nishiwaki, A., Yamada, T., Stewart, J. R., Deuter, M., & Sacks, E. J. (2019). Winter hardiness of Miscanthus (I): Overwintering ability and yield of new *Miscanthus × giganteus* genotypes in Illinois and Arkansas. *GCB Bioenergy, 11*(5), 691–705. https://doi.org/10.1111/gcbb.12588.

Doronin, V. A., Dryha, V. V., Kravchenko, Y. A., Mykolaiko, V. P., Karpuk, L. M., & Krasnoshtan, I. V. (2019). Growing of *Miscanthus × giganteus* planting material in the conditions of unstable moistening. *Eurasian Journal of Biosciences, 13*(2), 1101–1108. http://www.ejobios.org/article/growing-of-miscantus-giganteus-planting-material-in-the-conditions-of-unstable-moistening-7231.

Dubis, B., Jankowski, K. J., Załuski, D., & Sokólski, M. (2020). The effect of sewage sludge fertilization on the biomass yield of giant miscanthus and the energy balance of the production process. *Energy, 206,* 118189. https://doi.org/10.1016/j.energy.2020.118189.

Elliott, D. W., Lien, H.-L., & Zhang, W.-X. (2009). Degradation of lindane by zero-valent iron nanoparticles. *Journal of Environmental Engineering, 135*(5), 317–324. https://doi.org/10.1061/(asce)0733-9372(2009)135:5(317).

El-Temsah, Y. S., Sevcu, A., Bobcikova, K., Cernik, M., & Joner, E. J. (2016). DDT degradation efficiency and ecotoxicological effects of two types of nano-sized zero-valent iron (nZVI) in water and soil. *Chemosphere, 144,* 2221–2228. https://doi.org/10.1016/j.chemosphere.2015.10.122.

Evanylo, G. K. (2006). Chapter 10. Land application of biosolids. In K. C. Haering & G. K. Evanylo (Eds.), *The Mid-Atlantic Nutrient Management Handbook* (p.27). Virginia Cooperative Extension, Blacksburg, VA. https://vtechworks.lib.vt.edu/handle/10919/99502.

Faria, W. M., de Figueiredo, C. C., Coser, T. R., Vale, A. T., & Schneider, B. G. (2018). Is sewage sludge biochar capable of replacing inorganic fertilizers for corn production? Evidence from a two-year field experiment. *Archives of Agronomy and Soil Science, 64*(4), 505–519. https://doi.org/10.1080/03650340.2017.1360488.

George, E. F., Hall, M. A., & Klerk, G.-J. D. (2008). Plant growth regulators I: Introduction; Auxins, their analogues and inhibitors. In E. F. George, M. A. Hall, & G.-J. D. Klerk (Eds.), *Plant Propagation by Tissue Culture* (3rd ed., Vol. 1, pp.175–204). Springer, Netherlands. https://doi.org/10.1007/978-1-4020-5005-3_5.

Gonzalez, M., Miglioranza, K. S. B., Aizpún, J. E., Isla, F. I., & Peña, A. (2010). Assessing pesticide leaching and desorption in soils with different agricultural activities from Argentina (Pampa and Patagonia). *Chemosphere, 81*(3), 351–358. https://doi.org/10.1016/j.chemosphere.2010.07.021.

Heaton, E. A., Dohleman, F. G., Miguez, A. F., Juvik, J. A., Lozovaya, V., Widholm, J., Zabotina, O. A., McIsaac, G. F., David, M. B., Voigt, T. B., Boersma, N. N., & Long, S. P. (2010). Chapter 3: Miscanthus: A promising biomass crop. In J.-C. Kader & M. Delseny (Eds.), *Advances in Botanical Research* (Vol. 56, pp. 75–137). Academic Press, London, UK. https://doi.org/10.1016/B978-0-12-381518-7.00003-0.

Hu, B., Jarosch, A.-M., Gauder, M., Graeff-Hönninger, S., Schnitzler, J.-P., Grote, R., Rennenberg, H., & Kreuzwieser, J. (2018). VOC emissions and carbon balance of two bioenergy plantations in response to nitrogen fertilization: A comparison of Miscanthus and Salix. *Environmental Pollution, 237,* 205–217. https://doi.org/10.1016/j.envpol.2018.02.034.

Hülster, A., & Marschner, H. (1995). PCDD/PCDF-complexing compounds in zucchini. *Organohalogen Compounds, 24*, 493–496.

Humentik, M., Radejko, B., Fuchilo, Ya., Sinchenko, V., Ganzhenko, O., Bondar, V., Fursa, A., Kvak, V., Kharitonov, M., & Katelevskyi, V. (2018). *Production of Bioenergy Crops* (M. Humentik, Ed.). Komprint.

Inui, H., Sawada, M., Goto, J., Yamazaki, K., Kodama, N., Tsuruta, H., & Eun, H. (2013). A major latex-like protein is a key factor in crop contamination by persistent organic pollutants. *Plant Physiology, 161*(4), 2128–2135. https://doi.org/10.1104/pp.112.213645.

Iqbal, Y., Kiesel, A., Wagner, M., Nunn, C., Kalinina, O., Hastings, A. F. S. J., Clifton-Brown, J. C., & Lewandowski, I. (2017). Harvest time optimization for combustion quality of different Miscanthus genotypes across Europe. *Frontiers in Plant Science, 8*, 727. https://doi.org/10.3389/fpls.2017.00727.

Israr, M., Jewell, A., Kumar, D., & Sahi, S. V. (2011). Interactive effects of lead, copper, nickel and zinc on growth, metal uptake and antioxidative metabolism of *Sesbania drummondii. Journal of Hazardous Materials, 186*(2), 1520–1526. https://doi.org/10.1016/j.jhazmat.2010.12.021.

Jeżowski, S., Mos, M., Buckby, S., Cerazy-Waliszewska, J., Owczarzak, W., Mocek, A., Kaczmarek, Z., & McCalmont, J. P. (2017). Establishment, growth, and yield potential of the perennial grass *Miscanthus × giganteus* on degraded coal mine soils. *Frontiers in Plant Science, 8*, 726. https://doi.org/10.3389/fpls.2017.00726.

Kabata-Pendias, A., & Pendias, H. (2001). *Trace Elements in Soils and Plants* (3rd ed.). CRC Press, Boca Raton, FL.

Kaiser, C. M., & Sacks, E. J. (2015). Cold-tolerance of Miscanthus seedlings and effects of spring and autumn frosts on mature clonally replicated cultivars. *Crop Science, 55*(5), 2401–2416. https://doi.org/10.2135/cropsci2014.10.0679.

Kalinina, O., Nunn, C., Sanderson, R., Hastings, A. F. S., van der Weijde, T., Özgüven, M., Tarakanov, I., Schüle, H., Trindade, L. M., Dolstra, O., Schwarz, K.-U., Iqbal, Y., Kiesel, A., Mos, M., Lewandowski, I., & Clifton-Brown, J. C. (2017). Extending Miscanthus cultivation with novel germplasm at six contrasting sites. *Frontiers in Plant Science, 8*, 563. https://doi.org/10.3389/fpls.2017.00563.

Kansas Corn Yield Contest Winners. (2019). https://kscorn.com/yield-2/.

Kering, M. K., Butler, T. J., Biermacher, J. T., & Guretzky, J. A. (2012). Biomass yield and nutrient removal rates of perennial grasses under nitrogen fertilization. *BioEnergy Research, 5*(1), 61–70. https://doi.org/10.1007/s12155-011-9167-x.

Kharytonov, M., Pidlisnyuk, V., Stefanovska, T., Babenko, M., Martynova, N., & Rula, I. (2019). The estimation of *Miscanthus × giganteus'* adaptive potential for cultivation on the mining and post-mining lands in Ukraine. *Environmental Science and Pollution Research, 26*(3), 2974–2986.

Kirchmann, H., Börjesson, G., Kätterer, T., & Cohen, Y. (2017). From agricultural use of sewage sludge to nutrient extraction: A soil science outlook. *Ambio, 46*(2), 143–154. https://doi.org/10.1007/s13280-016-0816-3.

Kołodziej, B., Antonkiewicz, J., & Sugier, D. (2016). *Miscanthus × giganteus* as a biomass feedstock grown on municipal sewage sludge. *Industrial Crops and Products, 81*, 72–82. https://doi.org/10.1016/j.indcrop.2015.11.052.

Korenblum, E., Dong, Y., Szymanski, J., Panda, S., Jozwiak, A., Massalha, H., Meir, S., Rogachev, I., & Aharoni, A. (2020). Rhizosphere microbiome mediates systemic root metabolite exudation by root-to-root signaling. *Proceedings of the National Academy of Sciences, 117*(7), 3874–3883. https://doi.org/10.1073/pnas.1912130117.

Kucharik, C. J., VanLoocke, A., Lenters, J. D., & Motew, M. M. (2013). Miscanthus establishment and overwintering in the midwest USA: A regional modeling study of crop residue management on critical minimum soil temperatures. *PLoS One*, *8*(7), e68847. https://doi.org/10.1371/journal.pone.0068847.

Kucharski, R., Sas-Nowosielska, A., Małkowski, E., Japenga, J., Kuperberg, J. M., Pogrzeba, M., & Krzyżak, J. (2005). The use of indigenous plant species and calcium phosphate for the stabilization of highly metal-polluted sites in southern Poland. *Plant and Soil*, *273*(1), 291–305. https://doi.org/10.1007/s11104-004-8068-6.

Lee, D. K., Aberle, E., Anderson, E. K., and 55 others (2018). Biomass production of herbaceous energy crops in the United States: Field trial results and yield potential maps from the multiyear regional feedstock partnership. *GCB Bioenergy*, *10*(10), 698–716. https://doi.org/10.1111/gcbb.12493.

Lee, M.-S., Wycislo, A., Guo, J., Lee, D. K., & Voigt, T. (2017). Nitrogen fertilization effects on biomass production and yield components of *Miscanthus × giganteus*. *Frontiers in Plant Science*, *8*, 544. https://doi.org/10.3389/fpls.2017.00544.

Lehmann, J., Gaunt, J., & Rondon, M. (2006). Bio-char sequestration in terrestrial ecosystems – A review. *Mitigation and Adaptation Strategies for Global Change*, *11*(2), 403–427. https://doi.org/10.1007/s11027-005-9006-5.

Lehmann, J., Pereira da Silva, J., Steiner, C., Nehls, T., Zech, W., & Glaser, B. (2003). Nutrient availability and leaching in an archaeological Anthrosol and a Ferralsol of the Central Amazon basin: Fertilizer, manure and charcoal amendments. *Plant and Soil*, *249*(2), 343–357. https://doi.org/10.1023/a:1022833116184.

Liphadzi, M. S., Kirkham, M. B., & Paulsen, G. M. (2006). Auxin-enhanced root growth for phytoremediation of sewage-sludge amended soil. *Environmental Technology*, *27*(6), 695–704. https://doi.org/10.1080/09593332708618683.

Mamirova, A., Pidlisnyuk, V., Amirbekov, A., Ševců, A., Nurzhanova, A. (2020). Phytoremediation potential of *Miscanthus sinensis* And. in organochlorine pesticides contaminated soil amended by Tween 20 and Activated carbon. *Environmental Science and Pollution Research*. https://doi.org/10.1007/s11356-020-11609-y.

Mantineo, M., D'Agosta, G. M., Copani, V., Patanè, C., & Cosentino, S. L. (2009). Biomass yield and energy balance of three perennial crops for energy use in the semi-arid Mediterranean environment. *Field Crops Research*, *114*(2), 204–213. https://doi.org/10.1016/j.fcr.2009.07.020.

Matsuoka, S., Kennedy, A. J., dos Santos, E. G. D., Tomazela, A. L., & Rubio, L. C. S. (2014). Energy cane: Its concept, development, characteristics, and prospects. *Advances in Botany*, *2014*, 597275. https://doi.org/10.1155/2014/597275.

Maughan, M., Bollero, G., Lee, D. K., Darmody, R., Bonos, S., Cortese, L., Murphy, J., Gaussoin, R., Sousek, M., Williams, D., Williams, L., Miguez, F., & Voigt, T. (2012). *Miscanthus × giganteus* productivity: The effects of management in different environments. *GCB Bioenergy*, *4*(3), 253–265. https://doi.org/10.1111/j.1757-1707.2011.01144.x.

Mitros, T., Session, A. M., James, B. T., Wu, G. A., Belaffif, M. B., Clark, L. V., Shu, S., Dong, H., Barling, A., Holmes, J. R., Mattick, J. E., Bredeson, J. V., Liu, S., Farrar, K., Głowacka, K., Jeżowski, S., Barry, K., Chae, W. B., Juvik, J. A., … Rokhsar, D. S. (2020). Genome biology of the paleotetraploid perennial biomass crop Miscanthus. *Nature Communications*, *11*(1), 5442. https://doi.org/10.1038/s41467-020-18923-6.

Moberly Monitor. (2017). *Miscanthus offers Missouri farmers use for marginal land*. https://www.moberlymonitor.com/news/20170502/Miscanthus-offers-missouri-farmers-use-for-marginal-land.

Moyer, J. (2017). Adaptability of Miscanthus cultivars for biomass production. *Kansas Agricultural Experiment Station Research Reports*, 3(2). https://doi.org/10.4148/2378-5977.1377.

National Corn Growers Association. (2019). *National corn yield contest*. https://www.ncga.com/get-involved/national-corn-yield-contest.

Nebeská, D., Pidlisnyuk, V., Stefanovska, T., Trögl, J., Shapoval, P., Popelka, J., Cerný, J., Medkow, A., Kvak, V., & Malinská, H. (2019). Impact of plant growth regulators and soil properties on Miscanthus × giganteus biomass parameters and uptake of metals in military soils. *Reviews on Environmental Health*, 34(3), 283–291. https://doi.org/10.1515/reveh-2018-0088.

Nielsen, P. (1990). Elefantengrassanbau in Dänemark-Praktikerbericht. *Pflug Und Spaten*, 3, 1–4.

Ouattara, M. S., Laurent, A., Barbu, C., Berthou, M., Borujerdi, E., Butier, A., Malvoisin, P., Romelot, D., & Loyce, C. (2020). Effects of several establishment modes of Miscanthus × giganteus and Miscanthus sinensis on yields and yield trends. *GCB Bioenergy*, 12(7), 524–538. https://doi.org/10.1111/gcbb.12692.

Park, J.-W., & Boyd, S. A. (1999). Sorption of Chlorobiphenyls in sediment—Water systems containing nonionic surfactants. *Journal of Environmental Quality*, 28(3), 945–952. https://doi.org/10.2134/jeq1999.00472425002800030027x.

Pidlisnyuk, V., & Stefanovska, T. (2018). *Methods for growing M. × giganteus at the abandoned land* (Patent No. 127487), Ukraine.

Pidlisnyuk, V., Stefanovska, T., Lewis, E. E., Erickson, L. E., & Davis, L. C. (2014). Miscanthus as a productive biofuel crop for phytoremediation. *Critical Reviews in Plant Sciences*, 33(1), 1–19. https://doi.org/10.1080/07352689.2014.847616.

Pillai, H. P. S., & Kottekottil, J. (2016). Nano-phytotechnological remediation of endosulfan using zero valent iron nanoparticles. *Journal of Environmental Protection*, 7(5), 734. https://doi.org/10.4236/jep.2016.75066.

Pompeiano, A., Vita, F., Miele, S., & Guglielminetti, L. (2015). Freeze tolerance and physiological changes during cold acclimation of giant reed [*Arundo donax* (L.)]. *Grass and Forage Science*, 70(1), 168–175. https://doi.org/10.1111/gfs.12097.

Ponomarenko, S. P., Hrytsaenko, Z. M., & Tsygankova, V. A. (2013). Increase of plant resistance to diseases, pests and stresses with new biostimulants. *Acta Horticulturae*, 1009, 225–233. https://doi.org/10.17660/ActaHortic.2013.1009.27.

Ponomarenko, S. P., Stefanovska, T. R., Medkow, A. I., & Kapriy, M. (2017). Bioregulators of plant development in growing biofuel crops. *17th International Scientific Conference Sakharov Readings: Environmental Problems of the XXI Century*, II, 40–42.

Ponomarenko, S. P., Terek, O. I., Grytsaenko, Z. M., Babayants, O. V., Moiseeva, T. V., & Wenxiu, H. (2010). Bioregulation of growth and development of plants: Plant growth regulators in crop science. In S. P. Ponomarenko & H. O. Lutynska (Eds.), *Bioregulation of Microbial-Plant Systems* (pp. 251–291). Nichlava. Publishing Co., Kyiv, Ukraine, 472 pp. (in Russian).

Procházka, S., & Šebánek, J. (1997). *Plant Growth Regulators*. Academia Publishing, Prague, Czech Republic. 395 pp., ISBN 80-200-0597-8. (in Czech).

Pude, R. (2008). *Aktuelle Informationen aus der Miscanthus-Forschung*. www.miscanthus.de.

Pyter, R. J., Dohleman, F. G., & Voigt, T. B. (2010). Effects of rhizome size, depth of planting and cold storage on Miscanthus × giganteus establishment in the Midwestern USA. *Biomass and Bioenergy*, 34(10), 1466–1470. https://doi.org/10.1016/j.biombioe.2010.04.014.

Richter, G. M., Riche, A. B., Dailey, A. G., Gezan, S. A., & Powlson, D. S. (2008). Is UK biofuel supply from Miscanthus water-limited? *Soil Use and Management*, 24(3), 235–245. https://doi.org/10.1111/j.1475-2743.2008.00156.x.

Roik, M., Sinchenko, V., Purkin, V., Kvak, V., & Humentik, M. (Eds.). (2019). *Miscanthus in Ukraine*. FOP Yamchinskiy Press, Kyiv, Ukraine.

Sage, R. F., de Melo Peixoto, M., Friesen, P., & Deen, B. (2015). C4 bioenergy crops for cool climates, with special emphasis on perennial C4 grasses. *Journal of Experimental Botany*, 66(14), 4195–4212. https://doi.org/10.1093/jxb/erv123.

Scagline, S., Skousen, J., & Griggs, T. (2015). Switchgrass and miscanthus yields on reclaimed surface mines for bioenergy production. *Journal American Society of Mining and Reclamation*, 4, 80–90. https://doi.org/10.21000/jasmr15020080.

Schaap, M., Banzhaf, S., Scheuschner, T., Geupel, M., Hendriks, C., Kranenburg, R., Nagel, H.-D., Segers, A. J., von Schlutow, A., & Wichink Kruit, R. (2017). Atmospheric nitrogen deposition to terrestrial ecosystems across Germany. *Biogeosciences Discussions*, 1–24.

Skousen, J., Keene, T., Marra, M., & Gutta, B. (2013). Reclamation of mined land with switchgrass, Miscanthus, and Arundo for biofuel production. *Journal American Society of Mining and Reclamation*, 2, 177–191. https://doi.org/10.21000/jasmr13010160.

Smith, L. L., Allen, D. J., & Barney, J. N. (2015a). Yield potential and stand establishment for 20 candidate bioenergy feedstocks. *Biomass and Bioenergy*, 73, 145–154. https://doi.org/10.1016/j.biombioe.2014.12.015.

Smith, L. L., Askew, S. D., Hagood, E. S., & Barney, J. N. (2015b). Screening pre-emergence and postemergence herbicides for safety in bioenergy crops. *Weed Technology*, 29(1), 135–146. https://doi.org/10.1614/wt-d-14-00100.1.

Song, J.-S., Lim, S.-H., Lim, Y., Nah, G., Lee, D., & Kim, D.-S. (2016). Herbicide-based weed management in *Miscanthus sacchariflorus*. *BioEnergy Research*, 9(1), 326–334. https://doi.org/10.1007/s12155-015-9693-z.

Stewart, J. R., Toma, Y., Fernández, F. G., Nishiwaki, A., Yamada, T., & Bollero, G. (2009). The ecology and agronomy of *Miscanthus sinensis*, a species important to bioenergy crop development, in its native range in Japan: A review. *GCB Bioenergy*, 1(2), 126–153. https://doi.org/10.1111/j.1757-1707.2009.01010.x.

Tabak, M., Lisowska, A., Filipek-Mazur, B., & Antonkiewicz, J. (2020). The effect of amending soil with waste elemental sulfur on the availability of selected macroelements and heavy metals. *Processes*, 8(10), 1245. https://doi.org/10.3390/pr8101245.

Tian, H., Li, J., Mu, Z., Li, L., & Hao, Z. (2009). Effect of pH on DDT degradation in aqueous solution using bimetallic Ni/Fe nanoparticles. *Separation and Purification Technology*, 66(1), 84–89. https://doi.org/10.1016/j.seppur.2008.11.018.

Tomlinson, S., Carnell, E., Dore, A., Tipping, E., Misselbrook, T., Sutton, M., & Dragosits, U. (2011). *Historic nitrogen deposition: Long term trends* [poster]. http://www.ltls.org.uk/sites/default/files/Historic%20nitrogen%20deposition_Tomlinson.pdf.

Tsygankova, V. A., Andrusevych, Y. V., Babayants, O. V., Ponomarenko, S. P., Medkov, A. I., & Galkin, A. P. (2013b) Stimulation of plant immune protection against pathogenic fungi, pests and nematodes with growth regulators. *Physiology and Biochemistry of Cultivated Plants*, 45(2), 138–147 (in Ukrainian). https://www. academia.edu/4562192

Tsygankova, V. A., Yemets, A. I., Ponomarenko, S. P., Matvieieva, N. A., Chapkevich, S. E., & Kuchuk, N. V. (2013a). Increase in the synthesis of polyfructan in the cultures of chicory "hairy roots" with plant natural growth regulators. *International Journal of Biomedicine*, 3(2), 139–144. https://www.elibrary.ru/item.asp?id=20250613.

University of Iowa Facilities Management. (2020). *Renewable energy*. https://www.facilities.uiowa.edu/energy-environment/renewable-energy.

USDA/NRCS. (2011). *Planting and managing giant Miscanthus as a bioenergy crop* (Technical Note No. 4). USDA Natural Resources Conservation Service Plant Materials Program. https://www.nrcs.usda.gov/Internet/FSE_DOCUMENTS/stelprdb1044768.pdf.

U.S. Environmental Protection Agency. (1994). *A plain English guide to the EPA part 503 Biosolids Rule* (EPA/832/R-93/003). Office of Wastewater Management, United States Environmental Protection Agency.

U.S. Environmental Protection Agency. (2019). *Konza Prairie KNZ184*. https://www3.epa.gov/castnet/site_pages/KNZ184.html.

White, J. C., & Kottler, B. D. (2002). Citrate-mediated increase in the uptake of weathered 2,2-bis(p-chlorophenyl)1,1-dichloroethylene residues by plants. *Environmental Toxicology and Chemistry, 21*(3), 550–556. https://doi.org/10.1002/etc.5620210312.

White, J. C., Mattina, M. I., Lee, W.-Y., Eitzer, B. D., & Iannucci-Berger, W. (2003). Role of organic acids in enhancing the desorption and uptake of weathered p,p'-DDE by *Cucurbita pepo*. *Environmental Pollution, 124*(1), 71–80. https://doi.org/10.1016/S0269-7491(02)00409-8.

White, J. C., Parrish, Z. D., Gent, M. P. N., Iannucci-Berger, W., Eitzer, B. D., Isleyen, M., & Mattina, M. I. (2006). Soil amendments, plant age, and intercropping impact p,p'-DDE bioavailability to *Cucurbita pepo*. *Journal of Environmental Quality, 35*(4), 992–1000. https://doi.org/10.2134/jeq2005.0271.

Xue, S., Kalinina, O., & Lewandowski, I. (2015). Present and future options for Miscanthus propagation and establishment. *Renewable and Sustainable Energy Reviews, 49*, 1233–1246. https://doi.org/10.1016/j.rser.2015.04.168.

Zeng, F., Chen, S., Miao, Y., Wu, F., & Zhang, G. (2008). Changes of organic acid exudation and rhizosphere pH in rice plants under chromium stress. *Environmental Pollution, 155*(2), 284–289. https://doi.org/10.1016/j.envpol.2007.11.019.

Zinchenko, O. V. (2013). The evaluation of the effect of plants growth regulators on the photosynthesis intensity, survival rate and morphological indices of *Miscanthus giganteus*. *Scientific Notices of the Institute of Energy Crops and Sugar Beets, 19*, 47–51. http://www.bioenergy.gov.ua/sites/default/files/articles/47.pdf.

Zinchenko, O. V., Zinchenko, V. V., & Ponomarenko, S. P. (2016). Environmental aspects of the cultivation of Giant Miscanthus, potato and oat. In V. V. Pidlisnyuk & T. Stefanovska (Eds.), *Phytotechnology with Biomass Production for Re-cultivation of Lands Contaminated and Damaged by Military Activities* (pp. 102–105). Publisher house of NULES, Kyiv, Ukraine, ISBN 978-617-7396-14-6.

Zinchenko, V. O., Martynyuk, H. M., Zinchenko, O. V., Pitkevich, S., & Wisniewski, G. (2009). Features of growth of *Miscanthus × giganteus* under radioactive contamination. *The V Scientific Conference for Students and Young Scientists*, 138–140. Zhytomyr National Agroecological University, Zhytomyr, Ukraine (in Ukrainian).

6

Balancing Soil Health and Biomass Production

Larry E. Erickson and Kraig Roozeboom

Abstract

Soil is a living ecosystem with many different microorganisms (bacteria, fungi, and actinomycetes), microfauna (protists and nematodes), meso-fauna (arthropods, insects, mites, rotifera), and macrofauna (earthworms, termites, spiders, and isopods). Biodiversity is beneficial to soil health and improving biodiversity as part of a phytoremediation project is very desirable. Adding soil amendments such as compost or manure often improves soil health. Soil organic carbon is an important variable that provides carbon and energy for the organisms. At many sites with contaminants soil organic carbon is low at the start of the project. Soil health affects human health because the concentrations of elements and compounds in harvested fruits and vegetables depend on concentrations in the soil. If zinc concentration is low in a garden soil, there may be zinc deficiency in humans because of low concentrations in food from that garden. This chapter reviews aspects of soil quality and soil health and addresses the importance of improving soil health as part of a phytoremediation project.

CONTENTS

6.1 Introduction

Soil health, the capacity of soil to function as a vital living ecosystem, impacts the productivity of soil and ultimately human health. Reversing soil degradation by rebuilding healthy soils is a central theme of the emerging regenerative agriculture movement (Schreefel et al., 2020). Soils contaminated by military or other activities may require additional attention to address contamination and improve soil health. For example, biological organisms in the soil may have been reduced in number because of the contamination.

This chapter addresses the role of phytoremediation in improving soil health in the presence of contaminants. One goal at contaminated sites is to reduce risk of harmful exposure to contaminants, but a concurrent goal can be to improve the soil health sufficiently so the land can provide ecosystem services such as crop production. Soil health can be characterized based on physical, chemical, and biological properties.

Soil health and soil quality have both been used commonly. Soil health often includes the state of a population of living organisms in soil, while soil quality often refers to the physical and chemical state of the soil. Soil health is recognized as a local, regional, and global concern (Karlen & Rice, 2017). Education, research, and good practices are needed to improve soil quality. Government policies that are designed to improve soil health are beneficial.

6.2 Properties of Soils

The physical properties of soils include porosity, bulk density, water-holding capacity, texture, aggregation, infiltration, and penetration resistance. Chemical properties include concentrations of nutrients such as nitrogen, phosphorus, and potassium (N, P, and K), macronutrients such as calcium, micronutrients, such as zinc, organic carbon, pH, salinity, cation exchange capacity, and electrical conductivity (Brady & Weil, 2002). Biological properties include numbers of microorganisms, earthworms, other fauna, and soil respiration (Whalen & Sampedro, 2010).

Soil health is related to the measured values of soil properties, but contamination, availability of contaminants, and physical properties such as compaction affect soil health also. Actions to improve soil properties, such as adding amendments and building soil carbon, which is linked to both soil quality and soil health (Blanco-Canqui et al., 2015), can be beneficial to soil health and the phytoremediation process.

6.3 Soil Quality

In the world today, soil quality is poor in many locations. Low concentrations of nutrients and organic carbon result in lower yields. Salinity is an issue in locations where irrigation water has contributed to an increase in salt concentration. In this book, the emphasis is on applications of phytoremediation at contaminated sites. Many of these sites have physical issues because water-holding capacity is poor, porosity is low, and root penetration is difficult because of compaction. Both soil contamination and degraded soil quality must be addressed at many sites.

Soil tilth, the physical state of the soil for growing plants, often needs to be improved at many sites where phytoremediation is used to address contamination. The conditions at the soil surface and the conditions up to 20 cm below the surface are important for establishing vegetative cover. Good soil drainage and beneficial populations of bacteria, fungi, earthworms, nematodes, and other organisms are desirable (Chauhan & Mittu, 2015).

Assessment approaches of soil quality and indicators of soil health have been reviewed with respect to soil threats, ecosystem services, and functions (Bünemann et al., 2018). A total of 65 sets of indicators were reviewed; organic carbon, pH, available nitrogen, phosphorus and potassium, water storage, texture, and bulk density were included more than all other indicators. Biological indicators are important but are less commonly included in assessments because of the skills needed; however, including them can improve assessment results. The interpretation of assessment results requires experimental expertise or information on each test and its desired range of values. It is useful to be able to relate the ecosystem services to the values of the measured variables.

The United Nations Sustainable Development Goals include goal 2, which is zero hunger, and goal 15, which is life on land (sustainably manage forests, combat desertification, halt and reverse land degradation, and halt biodiversity loss). This goal 15 includes improving the health of ecosystems by addressing soil health to improve livelihoods. As population increases and quality of life improves, there is a great need to improve soil quality and the productivity of ecosystems that benefit from good soil health. As the effort to reduce greenhouse gas emissions advances and the transition to biodegradable products expands, land is needed for many purposes, and good soil health should become a priority for all countries.

The Food and Agriculture Organization (FAO) estimates that about one-third of soils that are used for agriculture have poor soil quality because of past use and management (FAO, 2011; Jian et al., 2020; Rodríguez-Eugenio et al., 2018). Low soil fertility, nutrient depletion, loss of soil from erosion, and low concentration of organic carbon are common reasons for poor soil quality. In a national survey in China, 16.1% of soil samples had measured values of one or more substances such as lead that exceeded the standards that are recommended for safety (Palansooriya et al., 2020).

Sustainable soil management for all soils is recognized as an important goal for society by FAO, and guidelines and principles have been developed (FAO, 2015, 2017). The guidelines include as follows: (i) minimize soil erosion; (ii) enhance soil organic matter content; (iii) foster soil nutrient balance and cycles; (iv) minimize soil salinization and alkalinization; (v) minimize soil contamination; (vi) minimize soil acidification; (vii) preserve and enhance soil biodiversity; (viii) minimize soil sealing; (ix) prevent and mitigate soil compaction; (x) Improve soil water management. Soil biodiversity and soil organic matter content are related because carbon, nutrients, and energy to support microorganisms and soil fauna are provided by the soil organic matter.

The principles in the soil charter recognize soil as a key resource that provides ecosystem services for food security. Sustainable soil management is based on knowledge of the physical, chemical, and biological state of the soil, and education about soils is beneficial. Global biodiversity of soil organisms is an important resource that needs to be sustained (FAO, 2015).

6.4 Soil Health Affects Human Health

Soil health has an impact on human health. The state of physical, mental, and social well-being is important for good human health (Brevik et al., 2020). The goal to create social and physical environments that contribute positively to good human health for all includes high-quality soil health that functions as a vital living ecosystem for growing plants that are beneficial to animal and human health. One example of soil health impacting human health is hypothyroidism/multinodular goiter because of soil deficiency in iodine. Another example is metal concentrations in foods that exceed the acceptable limits based on food safety (Brevik et al., 2020). Brevik et al. (2020) contains extensive tables on (i) properties of soils that may affect human health, such as zinc deficiency; (ii) persistent organic pollutants identified by the Stockholm Convention, such as chlordane; and (iii) human pathogens found in soil, such as Salmonella. Metals in soils are a global problem because of soil contamination associated with military activities, mining, industrial operations, and irrigation of crops with wastewater. The metals find their way into foods and are ingested, and inhalation of dust with metals is of concern. Pesticides are a major concern in many locations because about 25 million people working in agriculture are affected by pesticides each year (Brevik et al., 2020).

Phytoremediation is one of the better methods to improve soil health where pesticide-contaminated soil is found (Tarla et al., 2020). Micro- and macro-organisms in soil affect plant growth and health. Arbuscular mycorrhizal fungi have been found to enhance product yields and nutrient content. Plant growth promoting bacteria have been added to soil with beneficial results (Brevik et al., 2020; Pidlisnyuk et al., 2020). Endophytic bacteria and fungi produce enzymes that improve bioremediation of organic contaminants (Fagnano et al., 2020). Earthworms and nematodes are examples of macro-organisms that enhance nutrient cycling and diversity in soils. Earthworms improve soil structure and tilth. They are also impacted by pollutants and their population size is a measure of soil health.

6.5 Improving Soil Health Using Phytotechnology

Phytotechnology with biomass production can have multiple goals including (i) addressing contamination; (ii) improving soil health; (iii) improving biomass production of a useful product; (iv) adding soil carbon to improve soil quality and sequester carbon in the soil. Soil amendments may be added at contaminated sites because they are beneficial for the phytoremediation; however, it is desirable in selecting amendments to consider all four of the above goals. Soil amendments can impact pH, microbial populations, nutrient concentrations of N, P, K, and organic carbon, porosity, texture, salinity, and trace element concentrations.

A recent review by an international group of authors (Palansooriya et al., 2020) addressed soil amendments for soils containing potentially toxic elements. The authors present three valuable tables with information on a list of potentially toxic elements (As, Ba, Cd, Co, Cr, Hg, Ni, Mn, Mo, Pb, Sb, Se, V) including their chemistry in soils; organic soil amendments (animal waste, biochar, biosolids, compost, plant residues) that are beneficial in contaminated soils; and inorganic soil amendments (clay minerals, coal fly ash, industrial waste, liming materials, metal oxides, and phosphates) that have been used. The review includes information on many projects where soil amendments have been applied to reduce the availability of toxic elements.

Many organic soil amendments have beneficial value for soil health because the increase in soil organic matter improves soil structure, water-holding capacity, and nutrient availability. Biomass production is improved, microbial populations are larger, and the ecosystem functions better. The review includes a comprehensive discussion of research with biochar amendments in soils with toxic elements, including some information on 29 field studies reported by O'Connor et al. (2018). The yield with Miscanthus was increased using biochar in one of the studies. In general contaminant bioavailability was reduced by adding biochar, but the magnitude of the effect may decrease when pH decreases over time. Soil amendments have been reported to be cost-effective and beneficial to soil health and biomass production. When selecting amendments, it is important to evaluate their composition because toxic substances, salts, and other contaminants may be present. When using organic amendments at a new site where vegetation is being established, there may be a need to add soil fauna to enhance biodiversity.

In some applications of phytotechnology, there is a need to improve the biological state of the soil in order to improve soil health. The well-developed soil ecosystem includes about four trophic levels of organisms. Archaea, bacteria, fungi, actinomycetes, and algae provide ecosystem services by degrading organic compounds and making nutrients more available to plants. These are very small microorganisms of the order of 1 μm. Microfauna include protists (protozoa) and small nematodes that consume bacteria and other microorganisms. The microfauna are frequently larger than 2 μm and often less than 1 mm. There are soil mesofauna such as arthropods (insects), mites (acari), larger nematodes, and rotifera that are often in the size range between 0.1 and 5 mm. Macrofauna are about 1–50 mm in size and include earthworms, termites, spiders, and isopods (Whalen & Sampedro, 2010). Earthworms are very beneficial; they improve the texture of soil and nutrient cycling.

Although not conducted on a contaminated site, research comparing perennial crops such as miscanthus (*Miscanthus sacchariflorus*) and switchgrass (*Panicum virgatum* L.) with annual crops such as maize (*Zea mays* L.) and sorghum [*Sorghum bicolor* (L.) Moench] as potential cellulosic biofuel feedstocks illustrates the potential for achieving production and soil health goals simultaneously. Ethanol production potential of the perennial crops was less than that of the annual crops but still surpassed $3 \, m^3 ha^{-1} year^{-1}$ averaged over 10 years in the Central Great Plains of the US (Roozeboom et al., 2019). In that time, soil organic carbon increased in the 0–15 cm soil depth beneath perennial crops by 0.8–1.3 Mg C $ha^{-1}year^{-1}$ (McGowan et al., 2019). Greater soil organic carbon was associated with improvements in several parameters generally associated with greater soil health: root biomass, abundance of arbuscular mycorrhizae and saprophytic fungi, and soil aggregation, which is also associated with reduced soil loss.

6.6 Conclusions

Soil health and soil quality are very important because ecosystem services such as crop yields are greater when soil health is very good. In applications of phytoremediation with biomass production, it is beneficial to have multiple goals to reduce the effects of the contamination on soil health and to improve soil health and biomass productivity. Soil amendments that add organic carbon and living organisms may help to improve soil health, plant growth, and nutrient cycling. Education is generally of significant value because of the complexity of soil ecosystems and the many properties of soils that affect soil health and ecosystem services.

References

Blanco-Canqui, H., Shaver, T. M, Lindquist, J. L, Shapiro, C. A, Elmore, R. W., Francis, C. A., & Hergert, G. W. (2015). Cover crops and ecosystem services: Insights from studies in temperate soils. *Agronomy Journal, 107*, 2449–2474. https://doi.org/10.2134/agronj15.0086

Brady, N. C., & Weil, R. R. (2002). *The Nature and Properties of Soils* (13th ed.). Saddle River, NJ: Prentice Hall.

Brevik, E. C., Slaughter, L., Singh, B. R., Steffan, J. J., Collier, D., Barnhart, P., & Pereira, P. (2020). Soil and human health: Current status and future needs. *Air, Soil and Water Research, 13*, 1–23. https://doi.org/10.1177/1178622120934441

Bünemann, E. K., Bongiorno, G., Bai, Z., Creamer, R. E., De Deyn, G., de Goede, R., Fleskens, L., Geissen, V., Kuyper, T. W., Mäder, P., Pulleman, M., Sukkel, W., van Groenigen, J. W., & Brussaard, L. (2018). Soil quality – A critical review. *Soil Biology and Biochemistry, 120*, 105–125. https://doi.org/10.1016/j.soilbio.2018.01.030

Chauhan, A., & Mittu, B. (2015). Soil health-An issue of concern for environment and agriculture. *Journal of Bioremediation & Biodegredation, 6*(3), 1. https://doi.org/10.4172/2155-6199.1000286

Fagnano, M., Visconti, D., & Fiorentino, N. (2020). Agronomic approaches for characterization, remediation, and monitoring of contaminated sites. *Agronomy, 10*(9), 1335. https://doi.org/10.3390/agronomy10091335

FAO. (2011). *The state of the worlds land and water resources for food and agriculture*. Food and Agriculture Organization of the United Nations. http://fao.org.

FAO. (2015). *Revised world soil charter*. Food and Agriculture Organization of the United Nations. http://fao.org.

FAO. (2017). *Voluntary guidelines for sustainable soil management*. Food and Agriculture Organization of the United Nations. http://fao.org.

Jian, J., Du, X., & Stewart, R. D. (2020). A database for global soil health assessment. *Scientific Data, 7*(1), 16. https://doi.org/10.1038/s41597-020-0356-3.

Karlen, D. L., & Rice, C. W. (2017). *Enhancing Soil Health to Mitigate Soil Degradation*. Basel, Switzerland: MDPI. http://mdpi.com/books/pdfview/book/318.

McGowan, A. R., Nicoloso, R. S., Diop, H. E., Roozeboom, K. L., and Rice. C. W. (2019). Soil organic carbon, aggregation, and microbial community structure in annual and perennial biofuel crops. *Agronomy Journal, 111*, 128–142. https://doi.org/10.2134/agronj2018.04.0284.

O'Connor, D., Peng, T., Zhang, J., Tsang, D. C. W., Alessi, D. S., Shen, Z., Bolan, N. S., & Hou, D. (2018). Biochar application for the remediation of heavy metal polluted land: A review of in situ field trials. *Science of the Total Environment, 619–620*, 815–826. https://doi.org/10.1016/j.scitotenv.2017.11.132.

Palansooriya, K. N., Shaheen, S. M., Chen, S. S., Tsang, D. C. W., Hashimoto, Y., Hou, D., Bolan, N. S., Rinklebe, J., & Ok, Y. S. (2020). Soil amendments for immobilization of potentially toxic elements in contaminated soils: A critical review. *Environment International, 134*, 105046. https://doi.org/10.1016/j.envint.2019.105046.

Pidlisnyuk, V., Mamirova, A., Pranaw, K., Shapoval, P. Y., Trögl, J., & Nurzhanova, A. (2020). Potential role of plant growth-promoting bacteria in Miscanthus × *giganteus* phytotechnology applied to the trace elements contaminated soils. *International Biodeterioration & Biodegradation, 155*, 105103. https://doi.org/10.1016/j.ibiod.2020.105103.

Rodríguez-Eugenio, N., McLaughlin, M., & Pennock, D. (2018). *Soil pollution: A hidden reality.* Food and Agriculture Organization of the United Nations. http://fao.org.

Roozeboom, K. L., Wang, D., McGowan, A. R., Propheter, J. L., Staggenborg, S. A., & C. W. Rice. (2019). Long-term biomass and potential ethanol yields of annual and perennial biofuel crops. *Agronomy Journal, 111,* 74–83. https://doi.org/10.2134/agronj2018.03.0172.

Schreefel, L., Schulte, R. P. O., de Boer, I. J. M., Pas Schrijver, A., & van Zanten, H. H. E. (2020). Regenerative agriculture – The soil is the base. *Global Food Security,* 26, 100404, https://doi.org/10.1016/j.gfs.2020.100404.

Tarla, D. N., Erickson, L. E., Hettiarachchi, G. M., Amadi, S. I., Galkaduwa, M., Davis, L. C., Nurzhanova, A., & Pidlisnyuk, V. (2020). Phytoremediation and bioremediation of pesticide-contaminated soil. *Applied Sciences, 10*(4), 1217. https://doi.org/10.3390/app10041217.

Whalen, J. K., & Sampedro, L. (2010). *Soil Ecology and Management.* Vol. 8. São Paulo, Brazil: CABI.

7

Plant–Microbe Associations in Phytoremediation

Asil Nurzhanova, Aigerim Mamirova, Josef Trögl,
Diana Nebeská, and Valentina Pidlisnyuk

Abstract

Microorganisms are important partners with plants in phytotechnology applications. Plant–microbe relationships in phytoremediation include those of rhizobacteria which colonize root surfaces and biodegrade organic contaminants and other organic matter; endophytic bacteria that colonize the inner surface of plant stems and biodegrade organic compounds; and plant growth-promoting bacteria (PGPB) that have beneficial effects for plants. Plants produce organic substrates for the microbial populations. Because of root exudates, there are healthy numbers of bacteria near root surfaces that help with nutrient cycling and other ecosystem services. There is an emphasis on plant–microbe associations with Miscanthus; studies are conducted with and without PGPBs in soils contaminated with metals. Effects of PGPBs on bioconcentration factor and translocation factor are reported for Miscanthus growing in metal-contaminated soil.

CONTENTS

Microorganisms play key roles in soil ecosystem functions and supply important services, especially organic matter mineralization, nutrient cycling, and contribution to formation of humic substances (Blagodatskaya & Kuzyakov, 2013). Microorganisms associate with all parts of plant and form various interactions (Arora, 2013), and the term holobiont is used to stress the ubiquity of plant–microbe interactions (Hassani et al., 2018). The plant–microbe cooperation can be classified by various criteria, and the crucial benchmark is the profit which plant gains from such interactions which usually are neutral or positive (with mutualistic or symbiotic microorganisms); however, sometimes they could be negative (with parasitic or pathogenic microorganisms).

The symbiotic microorganisms can be classified as endosymbionts (living inside plant) or exosymbionts (living on the surface of plant). Symbiosis is a both-side positive relationship; nevertheless it is always described as "give cheap and take expensive", i.e., partners invest their surpluses and cover insufficiencies from the partner. Plants, as photoautotrophic organisms, mainly supply organic substances produced during photosynthesis. Indeed, microbial partners, often heterotrophic, consume them, provide to plants mineral nutrients (like fixed nitrogen or available phosphate), and protect against parasites or regulatory substances (Arora, 2013).

The effectiveness of phytoremediation depends on soil contamination level and historical background of soil exploitation. It also depends on the presence and accessibility of contaminants in the rhizosphere, their bioavailability to plant's root system, and the ability of plant–microbe association to intercept, absorb, accumulate, and/or degrade the contaminants (Vangronsveld et al., 2009). The plant–microbe associations are used to increase contaminant bioavailability and mobility in different environmental matrices (soil, water, wetlands, etc.) (Alkorta et al., 2004; Chu & Chan, 2003; Epelde et al., 2008; Mukherjee & Zimmerman, 2013). Contaminants inhibit plant growth and development and thereby reduce the phytoremediation effectiveness (Thion et al., 2013). In order to overcome this reduction, the plant–microbe partnership is used (Mitter et al., 2016). For example, the use of the C65 strain allowed *Populus euphratica* to more efficiently extract zinc from the contaminated environment, facilitating growth inhibition caused by heavy metals (Zhu et al., 2015).

7.1 Role of Plant–Microbe Association in Phytoremediation

The numerous studies are focused on using microorganisms (rhizo- and endophytic bacteria) for increasing the efficiency of phytoremediation technology and stimulating the plant development (Mitter et al., 2016). The main part of the research has been done in greenhouse conditions, while *in situ* field experiments have been rare (Glick, 2010; Guo & Chi, 2014; Kidd et al., 2017; Muratova et al., 2003a; Ren et al., 2019).

Rhizobacteria colonize the roots' surface while endophytic bacteria colonize the inner surface of stem without harming the plant, and both of the microorganisms are susceptible to biotic and abiotic factors. Another advantage of endophytes is that microorganisms are able to degrade the organic pollutants commonly widespread in endophytic populations. Endophytes can reduce the pollutants' phytotoxicity and the evaporation of volatile organic compounds (Shehzadi et al., 2014). Since endophytes are present inside the plant, they can interact more closely with the host plant in comparison with rhizobacteria. Before entering to plant, endophytes have to settle in rhizosphere and to attach to root surface. Organic compounds, i.e., root exudates, act as signals for the chemotactic movement of bacteria. During the transition from the plant rhizosphere to the endosphere, colonizing bacteria have to be able to quickly adapt to very different environment, i.e., pH, osmotic pressure, carbon source, and oxygen availability. They also have to overcome the plant's protective response to invasion, that is, the reactive oxygen species production, which causes stress for invasive bacteria. The important advantage of applying the endophytic degraders in phytoremediation technology is that any toxic xenobiotic absorbed by the plant can decompose inside the plant, thereby reducing the phytotoxic effect and eliminating any toxic effects on the herbivorous fauna living in or near contaminated sites (Ryan et al., 2008).

7.1.1 Endophytic Bacteria

Endophytic bacteria were first used to clean soil contaminated with organochloride herbicide 2,4-dichlorophenoxyacetic acid (Germaine et al., 2006). They reduced the accumulation of organic compounds in plant tissues and transpiration value. The improved degradation of pollutants was correlated with increasing the quantity of bacteria which were able to decompose pollutants in the plants. Woody plants: poplar and willow have been used to clean soil contaminated with various organic chemicals (Newman & Reynolds, 2005); inoculation of these plants with endophytic bacteria enhanced plant growth and degradation of various organic compounds. The partnership of different plant species and endophytes has been studied (Afzal et al., 2012; Germaine et al., 2006; Kang et al., 2012; Wang et al., 2010; Weyens et al., 2009; Yousaf et al., 2011), i.e., willow and *Burkholderia* sp. HU001, *Pseudomonas* sp. HU002 for Cd accumulation; poplar and *Enterobacter* sp. PDN3; multiflorum chaff (*L. multiflorum* var. Taurus) and *Pseudomonas sp.* ITRI53, *Pseudomonas* sp. MixRI75 with hydrocarbon degradation; multiflorum chaff (*L. multiflorum* var. Taurus), horned chaff (*L. corniculatus* var. Leo), alfalfa (*M. sativa* var. Harpe) and *Enterobacter ludwigii* with hydrocarbon degradation; poplar and *Pseudomonas putida* W619-TCE; lupine yellow and *Bacillus cepacia* VM1468; corn and wheat and *Burkholderia cepacia* FX2; peas (*Pisum sativum*) and *Pseudomonas putida*; wheat (*Triticum sp.*) and corn (*Z. mays*) and *Enterobacter sp.*

12J1; peas (*Pisum sativum*) and *Pseudomonas putida* VM1450 with the destruction of 2,4-dichlorophenoxyacetic acid. The potential of endophytic bacteria to enhance phytoremediation of soil contaminated with diesel fuel was compared with the potential of rhizobacteria (Afzal et al., 2012; Andria et al., 2009). Endophytes showed higher levels of colonization, especially in roots and shoots, and higher levels of expression of alkane-cleaving genes (alkB and CYP153) than the rhizobacterial strain. Inoculated endophytes colonize both the rhizosphere and the endosphere but are also metabolically active with respect to the decomposition of organic pollutants in soil and in plant itself (Afzal et al., 2014).

7.1.2 Rhizobacteria

The significant role of rhizobacteria in plant life is illustrated (Adesemoye & Kloepper, 2009; Afzal et al., 2014; Hayat et al., 2010; Mitter et al., 2016; Ortíz-Castro et al., 2009; Simpson et al., 2011). Active secretion by root cells of various substances, such as sugars, polysaccharides, amino acids, tricarboxylic acids, oxalic, malic, succinic and citric acids, fatty acids, sterols, phenols, enzymes and proteins, provides nutrients to microorganisms: inside root tissues and to root surfaces (rhizoplane) as well as to rhizosphere in close surroundings of roots (Döbereiner, 1989; Fan et al., 2018; Mitter et al., 2016). Plant exudates contribute to xenobiotic degradation through the following actions (Anderson & Coats, 1995):

- selective increasing in microorganisms destructors;
- increasing metabolism associated with growth;
- induction of joint metabolism with some microorganisms that have genes and plasmids responsible for degradation of pollutants.

Microorganisms secrete phytohormones, small molecules, or volatile compounds that can act directly or indirectly activate plant immunity or regulate plant growth and morphogenesis (Bhattacharyya & Jha, 2012; Glick, 2003; Ortíz-Castro et al., 2009). Indirect stimulation of plant growth is caused by reduction or preventing the harmful influence of phytotoxic compounds. It may include reducing iron prepared for phytopathogens in the rhizosphere; synthesis of enzymes that lyse fungal cell walls; competition with harmful microorganisms for their location on plant roots surface. Direct stimulation of plant growth by bacteria is caused by following factors (Dimkpa et al., 2008; Glick, 2003; Loper & Henkels, 1999; Rodríguez et al., 2006):

- providing plants with substances that are synthesized by bacteria (e.g., microbial nitrogen fixation);
- synthesis of siderophores, which can dissolve and accumulate iron from soil and supply it to plant cells;

- production of various phytohormones, including auxins (e.g., indole-3-acetic acid, cytokinins, and gibberellins) that affect growth;
- dissolution of minerals such as phosphorus which become readily available to plant;
- production of compounds which can influence growth and development, for example, 1-aminocyclopropane-1-carboxylic acid.

In the presence of pathogenic microorganisms, rhizosphere bacteria can synthesize various biocontrol agents (antibiotics, enzymes, siderophores, etc.) that suppress the growth of unwanted microbiota (Loper & Henkels, 1999; Mitter et al., 2016). Being developed on roots, the rhizosphere microbiomes affect plants by converting complex organic substances into accessible forms which stimulate growth and affect the morphology and physiology of plants; production of other specific metabolites, for example, ethylene, which causes early flowering (Glick, 2006).

Plant growth-promoting rhizobacteria (PGPR) are useful free-living rhizosphere bacteria that stimulate the plant's growth and are in association with them; they are found inside and around plant roots (Kloepper & Schroth, 1981; Belimov et al., 2005, 2009). These microorganisms are involved in complex ecological interactions in the rhizosphere, where they can influence plant health, growth, and stress response under unfavorable conditions, since some of them have a high destructive potential for pollutants (Glick, 2003; Hayat et al., 2010; Mitter et al., 2016).

PGPR have been widely used for improving crop development as biologically friendly fertilizers (Adesemoye & Kloepper, 2009; Ortíz-Castro et al., 2009; Simpson et al., 2011; Singh et al., 2011). Later the area of PGPRs application expanded, and they are considered for use in soil bioremediation, since many rhizosphere bacteria belonging to the PGPR group are resistant to pollutants (Costa et al., 2014). PGPRs are involved in processes of metal migration and have converted them to biologically available and soluble forms through effects of siderophores, organic acids, biosurfactants, biomethylation, and redox processes as part of the designed plant–microbial complexes (Ali et al., 2013; Belimov et al., 2005; Oh et al., 2015; Zhu et al., 2015). Effectiveness of phytoremediation depends on the activity of plant microbiome (Afzal et al., 2014; Weyens et al., 2009).

Campanella and Paul (2000) suggested that the root system of *Cucurbita pepo* secretes exudates of protein origin, which play an important role in the biodegradation of organochlorine compounds. Microbial degradation of pesticides mediated by enzyme systems is a promising approach for destruction of these toxic substances (Pascal-Lorber & Laurent, 2011) and assisted in the degradation of 4,4'-dichlorodiphenyl trichloroethane. Due to root exudates, the plant ensures the stable functioning of microorganisms, and they, in turn, contribute to the growth and development of the plant (Doornbos et al., 2012). Synergistic effects caused by interaction of plants and microorganisms stimulated the phytoremediation of soil contaminated by trace elements and organics: oil, polyaromatic hydrocarbons, polychlorinated biphenyls, organo-chlorine, nitroaromatic, and organophosphate compounds (Guo & Chi, 2014; Kidd et al., 2017; Muratova et al., 2003b). Through the inoculation of *Populus euphratica* by PGPR strain *Phyllobacterium sp.* C65, which produced auxin, assisted *Populus euphratica* to extract Zn more efficiently (Zhu et al., 2015).

7.2 Impact of PGPB Isolated from Contaminated Soil to Phytoremediation with Miscanthus

In order to increase the biomass harvest of Miscanthus while growing in con-taminated soils (Ben Fradj et al., 2020; Nsanganwimana et al., 2014; Pacheco-Torgal & Jalali, 2011), two main approaches can be used:

- soil treatment by different amendments such as fertilizers, sludge, biosolids, citric acid, Ethylenediamine tetraacedic acid (EDTA), and fungi (Antonkiewicz et al., 2019; Damodaran et al., 2013; Han et al., 2018; Hu et al., 2018; Alasmary, 2020);
- plant rhizome treatment by co-composting, plant growth regulators, and microorganisms (Khan et al., 2017; Leech et al., 2020; Nebeská et al., 2019).

The effectiveness of PGPB in the phytoremediation of metal contaminated soils can be explained by their ability to facilitate the adaptation of host plants to suboptimal soil conditions during stress state; promote plant growth; vary the bioavailability; relieve phytotoxicity in soil by producing amino acids, proteins, and antibiotics; and increase contaminant translocation within the plant (Oves et al., 2013). Also, PGPB can reduce the metals harmful effect by reduction, oxidation, methylation or de-methylation, compartmentalization, and conversion to a less toxic state (Hassan et al., 2017). Zeng et al. (2020) investigated the positive role of extracellular polymeric substances produced by *Bacillus* sp. S3 to detoxify different metals. Ndeddy Aka and Babalola (2016) showed that inoculation of soil by PGPB: *Pseudomonas aeruginosa* KP717554, *Alcaligenes faecalis* KP717561, and *Bacillus subtilis* KP717559 increased the amount of soluble Ni, Cd, and Cr in the soil by 51%, 50%, and 44%, respec-tively. Ma et al. (2015) researched the phytostabilization potential of PGPB in relation to metal contaminated soils: inoculation by *Pseudomonas* sp. A3R3 improved plant biomass production while *Psychrobacter* sp. SRS8 inocula-tion increased the accumulation of metals by plants. PGPB and *Miscanthus* sp. association in the metal contaminated soil was researched by Babu et al. (2015) and Schmidt et al. (2018). An endophytic PGPB *Pseudomonas koreensis* was explored for enhancing the production of *Miscanthus sinensis* growing in soil contaminated by As, Cd, Cu, Pb, and Zn (Babu et al., 2015).

Bacillus altitudinis KP-14 (*B. altitudinis* KP-14) is a plant growth-promoting bacterium isolated from postmining metal contaminated soil (Pranaw et al., 2020). The microbiological profile of *B. altitudinis* KP-14 is shown in Figure 7.1. The inoculation of *M. ×giganteus* rhizomes by this PGPB led to increasing of plant morphological parameters: aboveground (leaves and stems) and roots biomass increased by 23%, 86%, and 76%, respectively. In addition, artificially contaminated by Pb soil the increase in the root dry weight was detected.

The analysis of metal behavior in artificially contaminated by Pb soil is presented in Table 7.1. Increasing of Pb concentration in soil led to decreasing in Zn and Cu accumulation in roots. The inoculation of *M. ×giganteus* rhizomes resulted in increasing uptake of metals from the research soil: Mn (by 103%) > Zn (by 65%) > Sr, Pb (by 50%) > Cu (by 40%). These results can be explained by the ability of PGPB to mitigate metals tolerance (Babu et al., 2015; Schmidt et al., 2018). The metals tolerance mitigation process can be reached by their transformation into bioavailable and soluble form, organic acids, and siderophore production, diminishing phytotoxicity and altering the phytoavailability in contaminated soils (Ma et al., 2016).

The inoculation of *M. ×giganteus* rhizomes by the PGPB *B. altitudinis* strain KP-14 significantly enhanced plants' bioparameters and influenced the phytoremediation parameters: increased bioconcentration factor (BCF) and decreased translocation factor (TLF) while it was grown on the metal-contaminated soil. The increasing of BCF values can be explained by the rising of metals mobility; nevertheless, the phytoremediation process that

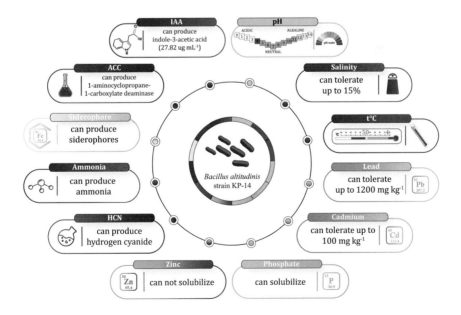

FIGURE 7.1
The tolerance profile and plant growth-promoting properties of *Bacillus altitudinis* strain KP-14. (Modified from Pidlisnyuk et al., 2020.)

TABLE 7.1

Trace Element Behavior in *M. × giganteus* Tissues during Growth in Artificially Contaminated by Pb Soil

Trace Element	Without *B. altitudinis* KP-14	With *B. altitudinis* KP-14
Pb	Mainly accumulated in roots.	Mainly accumulated in roots; translocation to aboveground biomass decreased.
Zn	Mainly accumulated in roots; the share accumulated in aboveground biomass increased from 13% to 20%.	The share accumulated in aboveground biomass increased from 16% to 22%; the share accumulated in roots decreased by 25.4% while the lead content in roots increased.
Cu	Mainly accumulated in roots; Accumulation in stems increased with increasing Pb concentrations in soil.	The share accumulated in stems remained almost the same (2%, 8%, and 4%) with increasing Pb concentration in soil. accumulation in roots increased by 58%–295%. total uptake increased as well (by 60%–238%).
Mn	Mainly accumulated in the aboveground biomass; the share in aboveground biomass (leaves in stems) was varied from 66% to 86%.	The total uptake increased, with increasing Pb concentration in soil, the effect decreased from 62% to 26%; accumulation in aboveground biomass decreased from 64% to 52%.
Sr	It evenly distributed throughout the plant; highest accumulation was in stems; the share in aboveground biomass varied from 50% to 69%.	It evenly distributed throughout the plant; highest accumulation was in the roots; accumulation in aboveground biomass (leaves and stems) reduced from 46% to 27%.

Source: Modified from Pidlisnyuk et al. (2020).

occurred could still be characterized as phytostabilization. The *M. × giganteus* root biomass increased with increasing Pb concentrations to tackle the Pb toxicity in the soil. With regard to the places of accumulation, the metals were divided into three groups: Pb, Zn, and Cu were predominantly accumulated in the roots; the Sr distribution was almost equal within the whole plant; and Mn predominantly accumulated in the leaves and stems.

7.3 Influence of Rhizobacteria Isolated from Miscanthus Rhizosphere to Phytoremediation of Trace Elements Contaminated Soil

The rhizosphere plays a significant role in phytoremediation of soil contaminated with trace elements (Jing et al., 2007). PGPRs are of interest among rhizosphere microorganisms (Shrivastava, 2017), because they facilitate the adaptation of host plants to suboptimal soil conditions during stressful conditions and increase the efficiency of phytoremediation (Hassan et al., 2017; Oves et al., 2013).

The PGPR strain *Agrobacterium* sp. Zn1-18 was used as an inoculant to increase the efficiency of phytoremediation of trace elements contaminated soil using *M. × giganteus* plants. The strain was isolated during the germination period from the rhizosphere of *M. × giganteus* when crop was cultivated in trace elements contaminated postindustrial soil (Table 7.2). Isolated strain showed PGPR properties: the ability to fix atmospheric nitrogen, to dissolve hard-to-reach phosphates, to synthesize siderophores, to produce phytohormone indolacetic acid (IAA), and ability to grow in medium that contains Zn ions up to 4.0 mmol L^{-1} and Pb ions up to 2.5 mmol L^{-1}.

Agrobacterium sp. Zn1-18 showed a significant effect on the bioparameters of *M. × giganteus* plants: the aboveground biomass of treated crop increased by 50%; the biomass of roots decreased by 77.3% on comparison with control untreated crop. Upon inoculation of *M. × giganteus* rhizome the roots' dry weight decreased by 34.8% while the dry weight of the aboveground biomass, on the contrary, increased by 7.1%. When comparing *M. × giganteus* grown on control and contaminated soils in the presence of PGPR, results revealed that the *Agrobacterium* sp. Zn1-18 increased the productivity of aboveground biomass by 7.1% while reduced the root dry weight. Thus, *Agrobacterium* sp. Zn1-18 promoted the growth and development of *M. × giganteus* on trace elements contaminated post-industrial soils.

TABLE 7.2

The Trace Element Concentrations in the Research Soils

Trace Element	MPC (mg kg⁻¹)	The Total Trace Element Concentration in Soil (mg kg⁻¹)	
		Control Soil	**Contaminated Soil**
V	150	50.5 ± 3.5	60.0 ± 12.0
Cr	6	35.0 ± 2.8	44.0 ± 7.1
Mn	1500	690.0 ± 56.5	810.0 ± 42.4
Fe	n/a	$19,600 \pm 424$	$21,450 \pm 2,757$
Co	5	14.5 ± 0.7	18.0 ± 1.4
Ni	4	32.0 ± 5.6	39.2 ± 2.8
Cu	3	31.0 ± 2.8	31.5 ± 7.7
Zn	55	101.0 ± 12.7	180.0 ± 14.1
As	2	11.5 ± 0.7	22.5 ± 4.9
Sr	7	66.0 ± 1.4	81.3 ± 0.8
Ba	0.1	125.0 ± 21.2	210.1 ± 28.2
Pb	32	20.0 ± 21.2	230.0 ± 44.2
U	n/a	0.8 ± 0.1	1.3 ± 0.2

Depth of soil sampling: 0–60 cm.
MPC, maximum permissible concentration.

The analysis of the abovementioned trace elements content in the roots and aboveground biomass of *M. ×giganteus*, when growing on contaminated and control soil with and without inoculation of *Agrobacterium* sp. Zn1-18, showed changes in the phytoremediation process (Figure 7.2). When crop grew in contaminated soil without inoculation of As, Pb, Co, Cr, Cu, V, and U accumulated mainly in the roots; Zn, Mn, and Sr—in the aboveground biomass; Ba and Ni accumulated uniformly in roots and aboveground biomass. When rhizomes were inoculated, phytoremediation potential of *M. ×giganteus* changed: content of Zn, Mn, As, Pb, Co, Cr, V, U, and Sr increased in roots, Ni accumulated in all plant parts, Cu accumulated mainly in the aboveground biomass, and for this trace elements accumulation increased by 41% compared to uninoculated system.

The calculated bioaccumulation factor and TLF coefficients (Nurzhanova et al., 2019) showed that inoculation of *M. ×giganteus* rhizomes with strain *Agrobacterium* sp. Zn1-18 increased the crop potential to extract As, Mn, Ba, Cu, Cr, Zn, and Sr from soil to aboveground biomass. The TLF values for Mn, Sr, Ba, Zn were ≥ 1, while for As, Cu, Pb, Co, Ni, V, Cr it was < 1 (Figure 7.2).

Overall using strain *Agrobacterium* sp. Zn1-18 reduced the total content of trace elements in plant tissues which may be due to decreasing of root dry weight which decreased adsorption surface and changed phytoremediation parameters.

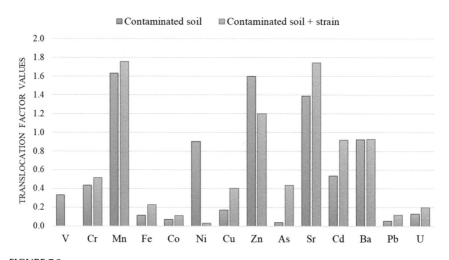

FIGURE 7.2
Influence of PGPB on the migration of elements from the *M. ×giganteus* root system to the aboveground biomass.

7.4 Changing of Soil Microbial Communities during Miscanthus Production at the Contaminated Military Land

The soil from the former military site (Sliač, Slovakia) contaminated by different trace elements was under investigation. *M. × giganteus* showed good growth in this soil during two vegetation seasons (Pidlisnyuk et al., 2018). The changes in soil microbial community, physiological state of soil microorganisms, and metabolic activities were investigated; rhizosphere and bulk soils were distinguished from each other (Nebeská et al., 2018).

During two-years growing period, the most significant changes could be attributed to time: comparison of first growing season vs. second growing season (Table 7.3). The indicators showed slight positive influence of *M. × giganteus* to soil and microbial community, which was documented by increased soil respiration, proportion of the fungal biomass, and decreased microbial stress indicators (*trans/cis* PLFA and *cy/pre* PLFA).

TABLE 7.3

Summary of the Changes of Soil Nutrients, Microbial Community Characteristics and Metabolic Activities during 2-Years Vegetation of *M. × giganteus* in Sliač Soil

Parameter	First Growing Season vs. Second Growing Season	Rhizosphere × Bulk Soil
Available P [mg kg^{-1} dwt]	0.27	1.02
Available K [mg kg^{-1} dwt]	1.90	0.00
Available Ca [mg kg^{-1} dwt]	0.03	0.32
Available Mg [mg kg^{-1} dwt]	1.72	2.82
S [%]	9.75↓ ***	2.15
N$_{tot}$ [%]	5.78↓ **	0.67
TOC [%]	0.14	0.08
Phosphatases [U g^{-1} dwt]	55.71↓ ***	2.52
Arylsulfatases [U g^{-1} dwt]	2.54	0.48
Proteases [U g^{-1} dwt]	12.42↑ ***	0.01
Oxidases [U g^{-1} dwt]	18.55↓ ***	0.51
Peroxidases [U g^{-1} dwt]	117.98↑ ***	0.16
Respiration [U g^{-1} dwt]	8.03↑ ***	0.94
PLFA$_{G+}$ [mg kg^{-1} dwt]	13.82↑ ***	0.00
PLFA$_{G-}$ [mg kg^{-1} dwt]	6.27↓ **	0.29
PLFA$_{Ac}$ [mg kg^{-1} dwt]	176.94↓ ***	0.58
PLFA$_{Fungi}$ [mg kg^{-1} dwt]	27.64↑ ***	0.98
trans/cis	2.92↓ *	3.27↑ *
cy/pre	82.77↓ ***	7.61↑ ***

Source: Modified from Nebeská et al. (2018).

Comparison was carried out by MANOVA: F-ratio statistic and its p-value; * $p < 0.1$, ** $p < 0.05$, *** $p < 0.01$; ↑ indicates a significant increase or ↓ a significant decrease.

The only negative change was decreasing the concentration of nutrients: S and N$_{tot}$, likely due to consumption by *M. × giganteus* and storage in rhizomes; other indicators did not change significantly, in particular, activities of extracellular enzymes.

The positive effect of *M. ×giganteus* on soil microorganisms can be registered by decreasing of stress indicators in rhizosphere soil in comparison to bulk soil.

References

Adesemoye, A., & Kloepper, J. (2009). Plant-microbes interactions in enhanced fertilizer-use efficiency. *Applied Microbiology and Biotechnology, 85*(1), 1–12. https://doi.org/10.1007/s00253-009-2196-0.

Afzal, M., Khan, Q., & Sessitsch, A. (2014). Endophytic bacteria: Prospects and applications for the phytoremediation of organic pollutants. *Chemosphere, 117*(1), 232–242. https://doi.org/10.1016/j.chemosphere.2014.06.078.

Afzal, M., Yousaf, S., Reichenauer, T., & Sessitsch, A. (2012). The inoculation method affects colonization and performance of bacterial inoculant strains in the phytoremediation of soil contaminated with diesel oil. *International Journal of Phytoremediation, 14*(1), 35–47. https://doi.org/10.1080/15226514.2011.552928.

Alasmary, Z. (2020). Laboratory-to field-scale investigations to evaluate phosphate amendments and Miscanthus for phytostabilization of lead-contaminated military sites, PhD Dissertation, Kansas State University, Manhattan. 297. https://hdl.handle.net/2097/40676.

Ali, H., Khan, E., & Sajad, M. (2013). Phytoremediation of heavy metals-Concepts and applications. *Chemosphere, 91*(7), 869–881. https://doi.org/10.1016/j.chemosphere.2013.01.075.

Alkorta, I., Hernández-Allica, J., Becerril, J., Amezaga, I., Albizu, I., & Garbisu, C. (2004). Recent findings on the phytoremediation of soils contaminated with environmentally toxic heavy metals and metalloids such as zinc, cadmium, lead, and arsenic. *Reviews in Environmental Science and Biotechnology, 3*(1), 71–90. https://doi.org/10.1023/B:RESB.0000040059.70899.3d.

Anderson, T., & Coats, J. (1995). Screening rhizosphere soil samples for the ability to mineralize elevated concentrations of atraz1ne and metolachlor. *Journal of Environmental Science and Health, Part B, 30*(4), 473–484. https://doi.org/10.1080/03601239509372948.

Andria, V., Reichenauer, T., & Sessitsch, A. (2009). Expression of alkane monooxygenase (alkB) genes by plant-associated bacteria in the rhizosphere and endosphere of Italian ryegrass (*Lolium multiflorum* L.) grown in diesel contaminated soil. *Environmental Pollution, 157*(12), 3347–3350. https://doi.org/10.1016/j.envpol.2009.08.023.

Antonkiewicz, J., Kołodziej, B., Bielińska, E., & Popławska, A. (2019). The possibility of using sewage sludge for energy crop cultivation exemplified by reed canary grass and giant miscanthus. *Soil Science Annual, 70*(1), 21–33. https://doi.org/10.2478/ssa-2019-0003.

Arora, N. K. (2013). *Plant Microbe Symbiosis: Fundamentals and Advances.* Springer, New Delhi.

Babu, A., Shea, P., Sudhakar, D., Jung, I., & Oh, B. (2015). Potential use of *Pseudomonas koreensis* AGB-1 in association with Miscanthus sinensis to remediate heavy metal(loid)-contaminated mining site soil. *Journal of Environmental Management, 151*, 160–166. https://doi.org/10.1016/j.jenvman.2014.12.045.

Belimov, A., Dodd, I., Hontzeas, N., Theobald, J., Safronova, V., & Davies, W. (2009). Rhizosphere bacteria containing 1-aminocyclopropane-1-carboxylate deaminase increase yield of plants grown in drying soil via both local and systemic hormone signalling. *New Phytologist, 181*(2), 413–423. https://doi.org/10.1111/j.1469-8137.2008.02657.x.

Belimov, A., Hontzeas, N., Safronova, V. I., Demchinskaya, S. V., Piluzza, G., Bullitta, S., & Glick, B. (2005). Cadmium-tolerant plant growth-promoting bacteria associated with the roots of Indian mustard (*Brassica juncea* L. Czern.). *Soil Biology and Biochemistry, 37*(2), 241–250. https://doi.org/10.1016/j.soilbio.2004.07.033.

Ben Fradj, N., Rozakis, S., Borzęcka, M., & Matyka, M. (2020). Miscanthus in the European bio-economy: A network analysis. *Industrial Crops and Products, 148*, 112281. https://doi.org/10.1016/j.indcrop.2020.112281.

Bhattacharyya, P., & Jha, D. (2012). Plant growth-promoting rhizobacteria (PGPR): Emergence in agriculture. *World Journal of Microbiology and Biotechnology, 28*(4), 1327–1350). https://doi.org/10.1007/s11274-011-0979-9.

Blagodatskaya, E., & Kuzyakov, Y. (2013). Active microorganisms in soil: Critical review of estimation criteria and approaches. *Soil Biology & Biochemistry, 67*, 192–211. https://doi.org/10.1016/j.soilbio.2013.08.024.

Campanella, B., & Paul, R. (2000). Presence, in the rhizosphere and leaf extracts of zucchini (*Cucurbita pepo* L.) and Melon (*Cucumis melo* L.), of Molecules, capable of increasing the apparent aqueous solubility of hydrophobic pollutants. *International Journal of Phytoremediation, 2*(2), 145–158. https://doi.org/10.1080/15226510008500036.

Chu, W., & Chan, K. (2003). The mechanism of the surfactant-aided soil washing system for hydrophobic and partial hydrophobic organics. *Science of the Total Environment, 307*(1–3), 83–92. https://doi.org/10.1016/S0048-9697(02)00461-8.

Costa, P., Granada, C., Ambrosini, A., Moreira, F., de Souza, R., dos Passos, J., Arruda, L., & Passaglia, L. (2014). A model to explain plant growth promotion traits: A multivariate analysis of 2,211 bacterial isolates. *PLoS One, 9*(12), e116020. https://doi.org/10.1371/journal.pone.0116020.

Damodaran, D., Vidya Shetty, K., & Raj Mohan, B. (2013). Effect of chelaters on bioaccumulation of Cd (II), Cu (II), Cr (VI), Pb (II) and Zn (II) in *Galerina vittiformis* from soil. *International Biodeterioration and Biodegradation, 85*, 182–188. https://doi.org/10.1016/j.ibiod.2013.05.031.

Dimkpa, C., Svatoš, A., Dabrowska, P., Schmidt, A., Boland, W., & Kothe, E. (2008). Involvement of siderophores in the reduction of metal-induced inhibition of auxin synthesis in *Streptomyces spp. Chemosphere, 74*(1), 19–25. https://doi.org/10.1016/j.chemosphere.2008.09.079.

Döbereiner, J. (1989). Isolation and identification of root associated diazotrophs. In Skinner, F.A., Boddey, R. M., & Fendrik, I. (eds.), *Nitrogen Fixation with Non-Legumes* (pp. 103–108). Springer, Dordrecht, Netherlands. https://doi.org/10.1007/978-94-009-0889-5_13.

Doornbos, R., Van Loon, L., & Bakker, P. (2012). Impact of root exudates and plant defense signaling on bacterial communities in the rhizosphere. A review. *Agronomy for Sustainable Development, 32*(1), 227–243. https://doi.org/10.1007/s13593-011-0028-y.

Epelde, L., Hernández-Allica, J., Becerril, J., Blanco, F., & Garbisu, C. (2008). Effects of chelates on plants and soil microbial community: Comparison of EDTA and EDDS for lead phytoextraction. *Science of the Total Environment, 401*(1–3), 21–28. https://doi.org/10.1016/j.scitotenv.2008.03.024.

Fan, D., Schwinghamer, T., & Smith, D. (2018). Isolation and diversity of culturable rhizobacteria associated with economically important crops and uncultivated plants in Québec, Canada. *Systematic and Applied Microbiology, 41*(6), 629–640. https://doi.org/10.1016/j.syapm.2018.06.004.

Germaine, K., Liu, X., Cabellos, G., Hogan, J., Ryan, D., & Dowling, D. (2006). Bacterial endophyte-enhanced phytoremediation of the organochlorine herbicide 2,4-dichlorophenoxyacetic acid. *FEMS Microbiology Ecology, 57*(2), 302–310. https://doi.org/10.1111/j.1574-6941.2006.00121.x.

Glick, B. (2003). Phytoremediation: Synergistic use of plants and bacteria to clean up the environment. *Biotechnology Advances, 21*(5), 383–393. https://doi.org/10.1016/S0734-9750(03)00055-7.

Glick, B. (2006). Modifying a plant's response to stress by decreasing ethylene production. In Mackova, M., Dowling, D., & Macek, T. (eds.), *Phytoremediation Rhizoremediation* (pp. 227–236). Springer Dordrecht, Netherlands. https://doi.org/10.1007/978-1-4020-4999-4_16.

Glick, B. (2010). Using soil bacteria to facilitate phytoremediation. *Biotechnology Advances, 28*(3), 367–374. https://doi.org/10.1016/j.biotechadv.2010.02.001.

Guo, J., & Chi, J. (2014). Effect of Cd-tolerant plant growth-promoting rhizobium on plant growth and Cd uptake by *Lolium multiflorum* Lam. and *Glycine max* (L.) Merr. in Cd-contaminated soil. *Plant and Soil, 375*(1–2), 205–214. https://doi.org/10.1007/s11104-013-1952-1.

Han, Y., Zhang, L., Gu, J., Zhao, J., & Fu, J. (2018). Citric acid and EDTA on the growth, photosynthetic properties and heavy metal accumulation of *Iris halophila* Pall. cultivated in Pb mine tailings. *International Biodeterioration and Biodegradation, 128*, 15–21. https://doi.org/10.1016/j.ibiod.2016.05.011.

Hassan, T., Bano, A., & Naz, I. (2017). Alleviation of heavy metals toxicity by the application of plant growth promoting rhizobacteria and effects on wheat grown in saline sodic field. *International Journal of Phytoremediation, 19*(6), 522–529. https://doi.org/10.1080/15226514.2016.1267696.

Hassani, M. A., Durán, P., & Hacquard, S. (2018). Microbial interactions within the plant holobiont. *Microbiome, 6*, 58. https://doi.org/10.1186/s40168-018-0445-0.

Hayat, R., Ali, S., Amara, U., Khalid, R., & Ahmed, I. (2010). Soil beneficial bacteria and their role in plant growth promotion: A review. *Annals of Microbiology, 60*(4), 579–598. https://doi.org/10.1007/s13213-010-0117-1.

Hu, B., Jarosch, A., Gauder, M., Graeff-Hönninger, S., Schnitzler, J., Grote, R., Rennenberg, H., & Kreuzwieser, J. (2018). VOC emissions and carbon balance of two bioenergy plantations in response to nitrogen fertilization: A comparison of *Miscanthus* and *Salix*. *Environmental Pollution, 237*, 205–217. https://doi.org/10.1016/j.envpol.2018.02.034.

Jing, Y., He, Z., & Yang, X. (2007). Role of soil rhizobacteria in phytoremediation of heavy metal contaminated soils. *Journal of Zhejiang University Science B, 8*(3), 192–207. https://doi.org/10.1631/jzus.2007.B0192.

Kang, J., Khan, Z., & Doty, S. (2012). Biodegradation of trichloroethylene by an endophyte of hybrid poplar. *Applied and Environmental Microbiology, 78*(9), 3504–3507. https://doi.org/10.1128/AEM.06852-11.

Khan, W., Ahmad, S., Yasin, N., Ali, A., & Ahmad, A. (2017). Effect of *Pseudomonas fluorescens* RB4 and *Bacillus subtilis* 189 on the phytoremediation potential of *Catharanthus roseus* (L.) in Cu and Pb-contaminated soils. *International Journal of Phytoremediation, 19*(6), 514–521. https://doi.org/10.1080/15226514. 2016.1254154.

Kidd, P., Álvarez-López, V., Becerra-Castro, C., Cabello-Conejo, M., & Prieto-Fernández, Á. (2017). Potential role of plant-associated bacteria in plant metal uptake and implications in phytotechnologies. *Advances in Botanical Research, 83*, 87–126. https://doi.org/10.1016/bs.abr.2016.12.004.

Kloepper, J., & Schroth, M. (1981). Relationship of in vitro antibiosis of plant growth-promoting rhizobacteria to plant growth and the displacement of root microflora. *Phytopathology, 71*(10), 1020. https://doi.org/10.1094/phyto-71-1020.

Leech, C., Tighe, M., Pereg, L., Winter, G., McMillan, M., Esmaeili, A., & Wilson, S. (2020). Bioaccessibility constrains the co-composting bioremediation of field aged PAH contaminated soils. *International Biodeterioration and Biodegradation, 149*, 104922. https://doi.org/10.1016/j.ibiod.2020.104922.

Loper, J., & Henkels, M. (1999). Utilization of heterologous siderophores enhances levels of iron available to *Pseudomonas putida* in the rhizosphere. *Applied and Environmental Microbiology, 65*(12), 5357–5363. https://doi.org/10.1128/ aem.65.12.5357-5363.1999.

Ma, Y., Rajkumar, M., Rocha, I., Oliveira, R., & Freitas, H. (2015). Serpentine bacteria influence metal translocation and bioconcentration of *Brassica juncea* and *Ricinus communis* grown in multi-metal polluted soils. *Frontiers in Plant Science, 5*(JAN), 757. https://doi.org/10.3389/fpls.2014.00757.

Ma, Y., Rajkumar, M., Zhang, C., & Freitas, H. (2016). Beneficial role of bacterial endophytes in heavy metal phytoremediation. *Journal of Environmental Management, 174*, 14–25. https://doi.org/10.1016/j.jenvman.2016.02.047.

Mitter, B., Pfaffenbichler, N., & Sessitsch, A. (2016). Plant–microbe partnerships in 2020. *Microbial Biotechnology, 9*(5), 635–640. https://doi. org/10.1111/1751-7915.12382.

Mukherjee, A., & Zimmerman, A. (2013). Organic carbon and nutrient release from a range of laboratory-produced biochars and biochar-soil mixtures. *Geoderma, 193–194*, 122–130. https://doi.org/10.1016/j.geoderma.2012.10.002.

Muratova, A., Hübner, T., Narula, N., Wand, H., Turkovskaya, O., Kuschk, P., Jahn, R., & Merbach, W. (2003a). Rhizosphere microflora of plants used for the phytoremediation of bitumen-contaminated soil. *Microbiological Research, 158*(2), 151–161. https://doi.org/10.1078/0944-5013-00187.

Muratova, A., Hübner, T., Tischer, S., Turkovskaya, O., Möder, M., & Kuschk, P. (2003b). Plant - Rhizosphere-microflora association during phytoremediation of PAH-contaminated soil. *International Journal of Phytoremediation, 5*(2), 137–151. https://doi.org/10.1080/713610176.

Ndeddy Aka, R., & Babalola, O. (2016). Effect of bacterial inoculation of strains of *Pseudomonas aeruginosa*, *Alcaligenes feacalis* and *Bacillus subtilis* on germination, growth and heavy metal (Cd, Cr, and Ni) uptake of *Brassica juncea*. *International Journal of Phytoremediation, 18*(2), 200–209. https://doi.org/10.1080/15226514.2015 .1073671.

Nebeská, D., Pidlisnyuk, V., Stefanovska, T., Trögl, J., Shapoval, P., Popelka, J., Cerný, J., Medkow, A., Kvak, V., & Malinská, H. (2019). Impact of plant growth regulators and soil properties on *Miscanthus x giganteus* biomass parameters and uptake of metals in military soils. *Reviews on Environmental Health, 34*(3), 283–291. https://doi.org/10.1515/reveh-2018-0088.

Nebeská, D., Trögl, J., Pidlisnyuk, V., Popelka, J., Dáňová, P. V., Usťak, S. & Honzík, R. (2018). Effect of growing *Miscanthus x giganteus* on soil microbial communities in post-military soil. *Sustainability, 10,* 4021. https://doi.org/10.3390/su10114021.

Newman, L., & Reynolds, C. (2005). Bacteria and phytoremediation: New uses for endophytic bacteria in plants. *Trends in Biotechnology, 23*(1), 6–8. https://doi.org/10.1016/j.tibtech.2004.11.010.

Nsanganwimana, F., Pourrut, B., Mench, M., & Douay, F. (2014). Suitability of Miscanthus species for managing inorganic and organic contaminated land and restoring ecosystem services. A review. *Journal of Environmental Management, 143,* 123–134. https://doi.org/10.1016/j.jenvman.2014.04.027.

Nurzhanova, A., Pidlisnyuk, V., Abit, K., Nurzhanov, C., Kenessov, B., Stefanovska, T., & Erickson, L. (2019). Comparative assessment of using *Miscanthus × giganteus* for remediation of soils contaminated by heavy metals: a case of military and mining sites. *Environmental Science and Pollution Research, 26*(13), 13320–13333. https://doi.org/10.1007/s11356-019-04707-z.

Oh, K., Cao, T., Cheng, H., Liang, X., Hu, X., Yan, L., Yonemochi, S., & Takahi, S. (2015). Phytoremediation potential of sorghum as a biofuel crop and the enhancement effects with microbe inoculation in heavy metal contaminated soil. *Journal of Biosciences and Medicines, 3*(6), 9–14. https://doi.org/10.4236/jbm.2015.36002.

Ortíz-Castro, R., Contreras-Cornejo, H., Macías-Rodríguez, L., & López-Bucio, J. (2009). The role of microbial signals in plant growth and development. *Plant Signaling & Behavior, 4*(8), 701–712. https://doi.org/10.4161/psb.4.8.9047.

Oves, M., Khan, M., & Zaidi, A. (2013). Chromium reducing and plant growth promoting novel strain *Pseudomonas aeruginosa* OSG41 enhance chickpea growth in chromium amended soils. *European Journal of Soil Biology, 56,* 72–83. https://doi.org/10.1016/j.ejsobi.2013.02.002.

Pacheco-Torgal, F., & Jalali, S. (2011). Cementitious building materials reinforced with vegetable fibres: A review. *Construction and Building Materials, 25*(2), 575–581. https://doi.org/10.1016/j.conbuildmat.2010.07.024.

Pascal-Lorber, S., & Laurent, F. (2011). Phytoremediation Techniques for Pesticide Contaminations. In Lichtfouse, E. (eds.), *Alternative Farming Systems, Biotechnology, Drought Stress and Ecological Fertilisation* (pp. 77–105). Springer, Dordrecht, Netherlands. https://doi.org/10.1007/978-94-007-0186-1_4.

Pidlisnyuk, V., Erickson, L. E., Trogl, J., Shapoval, P., Popelka, J., Davis, L., Stefanovska, T., & Hettiarachchi, G. (2018). Metals uptake behaviour in *Miscanthus × giganteus* plant during growth at the contamined soil from the military site in Sliac, Slovakia. *Polish Journal of Chemical Technology, 20*(2), 1–7. https://doi.org/10.2478/pjct-2018-0016.

Pidlisnyuk, V., Mamirova, A., Pranaw, K., Shapoval, P., Trögl, J., & Nurzhanova, A. (2020). Potential role of plant growth-promoting bacteria in *Miscanthus × giganteus* phytotechnology applied to the trace elements contaminated soils. *International Biodeterioration & Biodegradation, 155,* 105103. https://doi.org/10.1016/j.ibiod.2020.105103.

Pranaw, K., Pidlisnyuk, V., Trögl, J., & Malinská, H. (2020). Bioprospecting of a novel plant growth-promoting bacterium *Bacillus altitudinis* KP-14 for enhancing *Miscanthus ×giganteus* growth in metals contaminated soil. *Biology, 9*(9), 305. https://doi.org/10.3390/biology9090305.

Ren, X., Guo, S., Tian, W., Chen, Y., Han, H., Chen, E., Li, B., Li, Y., & Chen, Z. (2019). Effects of plant growth-promoting bacteria (PGPB) inoculation on the growth, antioxidant activity, Cu uptake, and bacterial community structure of rape (*Brassica napus* L.) grown in Cu-contaminated agricultural soil. *Frontiers in Microbiology, 10*(JUN), 1455. https://doi.org/10.3389/fmicb.2019.01455.

Rodríguez, H., Fraga, R., Gonzalez, T., & Bashan, Y. (2006). Genetics of phosphate solubilization and its potential applications for improving plant growth-promoting bacteria. *Plant and Soil, 287*(1–2), 15–21. https://doi.org/10.1007/s11104-006-9056-9.

Ryan, R., Germaine, K., Franks, A., Ryan, D., & Dowling, D. (2008). Bacterial endophytes: Recent developments and applications. *FEMS Microbiology Letters, 278*(1), 1–9. https://doi.org/10.1111/j.1574-6968.2007.00918.x.

Schmidt, C., Mrnka, L., Frantík, T., Lovecká, P., & Vosátka, M. (2018). Plant growth promotion of *Miscanthus ×giganteus* by endophytic bacteria and fungi on non-polluted and polluted soils. *World Journal of Microbiology and Biotechnology, 34*(3), 1–20. https://doi.org/10.1007/s11274-018-2426-7.

Shehzadi, M., Afzal, M., Khan, M., Islam, E., Mobin, A., Anwar, S., & Khan, Q. (2014). Enhanced degradation of textile effluent in constructed wetland system using *Typha domingensis* and textile effluent-degrading endophytic bacteria. *Water Research, 58*, 152–159. https://doi.org/10.1016/j.watres.2014.03.064.

Shrivastava, U. (2017). Molecular diversity assessment of plant growth promoting rhizobacteria using denaturing gradient gel electrophoresis (DGGE) of 16s rRNA gene. *International Journal of Applied Sciences and Biotechnology, 5*(1), 72–80. https://doi.org/10.3126/ijasbt.v5i1.17029.

Simpson, R., Oberson, A., Culvenor, R., Ryan, M., Veneklaas, E., Lambers, H., Lynch, J., Ryan, P., Delhaize, E., Smith, F., Smith, S., Harvey, P., & Richardson, A. (2011). Strategies and agronomic interventions to improve the phosphorus-use efficiency of farming systems. *Plant and Soil, 349*(1–2), 89–120. https://doi.org/10.1007/s11104-011-0880-1.

Singh, J., Pandey, V., & Singh, D. (2011). Efficient soil microorganisms: A new dimension for sustainable agriculture and environmental development. *Agriculture, Ecosystems and Environment, 140*(3–4), 339–353. https://doi.org/10.1016/j.agee.2011.01.017.

Thion, C., Cébron, A., Beguiristain, T., & Leyval, C. (2013). Inoculation of PAH-degrading strains of *Fusarium solani* and *Arthrobacter oxydans* in rhizospheric sand and soil microcosms: Microbial interactions and PAH dissipation. *Biodegradation, 24*(4), 569–581. https://doi.org/10.1007/s10532-013-9628-3.

Vangronsveld, J., Herzig, R., Weyens, N., Boulet, J., Adriaensen, K., Ruttens, A., Thewys, T., Vassilev, A., Meers, E., Nehnevajova, E., van der Lelie, D., & Mench, M. (2009). Phytoremediation of contaminated soils and groundwater: Lessons from the field. *Environmental Science and Pollution Research, 16*(7), 765–794. https://doi.org/10.1007/s11356-009-0213-6.

Wang, L., Wang, W., Lai, Q., & Shao, Z. (2010). Gene diversity of CYP153A and AlkB alkane hydroxylases in oil-degrading bacteria isolated from the Atlantic Ocean. *Environmental Microbiology, 12*(5), 1230–1242. https://doi.org/10.1111/j.1462-2920.2010.02165.x.

Weyens, N., van der Lelie, D., Taghavi, S., Newman, L., & Vangronsveld, J. (2009). Exploiting plant-microbe partnerships to improve biomass production and remediation. *Trends in Biotechnology, 27*(10), 591–598. https://doi.org/10.1016/j.tibtech.2009.07.006.

Yousaf, S., Afzal, M., Reichenauer, T. G., Brady, C. L., & Sessitsch, A. (2011). Hydrocarbon degradation, plant colonization and gene expression of alkane degradation genes by endophytic *Enterobacter ludwigii* strains. *Environmental Pollution, 159*(10), 2675–2683. https://doi.org/10.1016/j.envpol.2011.05.031.

Zeng, W., Li, F., Wu, C., Yu, R., Wu, X., Shen, L., Liu, Y., Qiu, G., & Li, J. (2020). Role of extracellular polymeric substance (EPS) in toxicity response of soil bacteria *Bacillus sp.* S3 to multiple heavy metals. *Bioprocess and Biosystems Engineering, 43*(1), 153–167. https://doi.org/10.1007/s00449-019-02213-7.

Zhu, D., Ouyang, L., Xu, Z., & Zhang, L. (2015). Rhizobacteria of *Populus euphratica* promoting plant growth against heavy metals. *International Journal of Phytoremediation, 17*(10), 973–980. https://doi.org/10.1080/15226514.2014.981242.

8

Plant Feeding Insects and Nematodes Associated with Miscanthus

Tatyana Stefanovska, Valentina Pidlisnyuk, and Andrzej Skwiercz

Abstract

Plant feeding insects and nematodes have the potential to impact Miscanthus growth and product yield. The biological ecosystems associated with Miscanthus include many organisms that are beneficial to Miscanthus; however, a number of pests have been identified and studied. Some plant feeding pests have importance for both Miscanthus and some food and feed crops. It is important to consider pest migration from one crop to another when fields are nearby. Miscanthus mealybugs, aphids, May beetles, nematodes, armyworms, and rootworms have been found in fields of Miscanthus. Norovirus and Tobrivirus have nematode vectors that may transmit these viruses. This chapter provides information on important plant feedings insects, nematodes, and other pests that have been found with Miscanthus and have been studied and reported in published literature.

CONTENTS

8.1 Introduction

Bioenergy systems aimed at replacing fossil fuels with bio-based resources contribute significantly to the shift in agricultural and marginal land use (Warner et al., 2013). The creation and/or conversion of land occupied by bioenergy systems can result in the loss of areas, currently providing additional ecosystem services (Cook, 1991; Stefanovska et al., 2015; Reid et al., 2020), including the loss of diversity of flora and fauna (Elshout et al., 2019). By reducing plant varieties in bioenergy crop plantations, insect and nematode species/functional diversity, community structure will be affected by modifying the associates' food webs (Landis & Werling, 2010). Bioenergy crops and adjacent to them food/feed crops can be affected by this reduction (Thomson & Hoffmann, 2011). Biofuel crops' changing may also influence the temporal and spatial distribution, and efficacy of beneficial organisms involving in pest control that would reduce or decrease the efficacy of biocontrol (Werling et al., 2011).

Miscanthus spp. produces high yield with low input (Gołąb-Bogacz et al., 2020) on agricultural as well as marginal, deteriorated, and contaminated land (Gruss et al., 2019) and determines the great socio-economic potential (Ben Fradj et al., 2020). *M. ×giganteus* is a habitat for many insects and nematodes, which can hold beneficial and plant-feeding status (Winkler et al., 2020). This chapter will be focused on plant feeding insects and nematodes that were reported on Miscanthus plantations in Europe and USA during different stages of crop development with explanation of their potential to reduce the crop yield.

8.2 Plant Feeding Insects with Piercing-Sucking Mouth Parts

8.2.1 Miscanthus Mealybug

M. ×giganteus parenteral plants *M. sacchariflorus* and *M. sinensis* are susceptible to several pests in native countries (Stefanovska et al., 2017a). Mealybug *Miscanthiococcus miscanthi* Takahashi (Hemiptera: Pseudococcidae) is a subtropical species (most often distributed within the boundaries of 23.24°–30.16° Northern Latitude); however, it can be spread to higher latitudes: 44.10°. Several studies show that in Southeastern states of the US, Miscanthus mealybugs are becoming a significant problem (Wheeler, 2013).

8.2.1.1 Identification

The body of mealybug adult females is wingless, soft, and oval, approximately 3 mm long. They are covered by white wax. Male mealybugs are up to 2 mm long with four eyes, two wings, and long tails. Newly hatched nymphs (crawlers) are flat, oval, and yellow, and generally do not have a waxy coating. Older nymphs are covered with fluffy, white wax.

8.2.1.2 Life Cycle

Adult females and crawlers overwinter and commonly emerge by May. Adult females of most mealybugs lay 100–200 or more eggs in cottony egg sacs that are attached to crowns, leaves, or twigs. It takes about a month for crawlers to mature. Spreading of Miscanthus mealybugs generally happens via propagating infested plants or blowing of crawlers into nearby plants. This pest has three generations per year.

8.2.1.3 Damage

The species suck sap from plant phloem, causing yellowing and twisting of leaves, stunted and slowing down of plant growth. Insects also can damage plants by excreting the sticky honeydew which may be a substance for black sooty mold fungus growth. The good signs of a possible infestation of Miscanthus mealybug are purple spots on infested stems.

Because *M. ×giganteus* is capable to form a strong root system and big vegetative mass, a thick layer of dead organic litter can serve as niche for warm climate mealybug *M. miscanthi*, and for various indigenous coccid species feeding on cereals (Kosztarab & Kozár, 2012). Moving to the root system under the cover of organic litter Miscanthus mealybug is capable to survive at air temperature which is much lower than in its natural habitat conditions (Stefanovska et al., 2017a).

8.2.2 Aphids

Twenty-one aphids are known to use Miscanthus (mostly *M. sinensis*) as a host; therefore, there is a potential for aphid damage of the crop. Results of field study carried out by Semere and Slater (2007) demonstrated domination of family *Aphidiidae* among *Homoptera*. There still is a gap in information about broad Aphid biodiversity at Miscanthus. The extensive field surveys carried out in USA (Bradshaw et al., 2010) and laboratory studies (Pallipparambil et al., 2014) indicated that among the insects reported in Miscanthus, the yellow sugarcane aphid, *Sipha flava* Forbes, is a potential pest of *M. ×giganteus*. *Sipha flava* (Homoptera: Aphididae) is native to North America (Hentz & Nuessly, 2002) and has been recorded on approximately 60 plant species, including *Cyperaceae, Poaceae*, and *Commelinaceae* (Kindler & Dalrymple, 1999).

8.2.2.1 Identification

Wingless yellow sugarcane aphids and nymphs are 1.3–2 mm in length bright yellow or green at the low temperature, and the insect body is tightly covered by hairs. Rows of spots are present down the top and along lateral margins of the abdomen. The spots' size and tailpipes are reduced compared to other aphid species.

8.2.2.2 Life Cycle

The species reproduce by parthenogenesis during the year in warmer climates, but sexual forms occur in regions with cold winters and it can overwinter in eggs. Development to adulthood takes 8–15 days and is highly dependent on temperature and host plants. This pest has many generations.

8.2.2.3 Damage

The yellow sugarcane aphid causes damage to sorghum, sugarcane, and several species of lawn pasture grass (Kindler & Dalrymple, 1999). Feeding initially results in leaves turning to yellow or red, depending on the host plant and temperature. Prolonged feeding can lead to premature senescence of leaves and plant or stalk death. Colonies or groups of sugarcane aphids are located around the midrib of the bottom side. Additionally, sugarcane aphids are prolific producers of honeydew which supports growth of sooty mold fungi. Yield reductions commonly occur due to feeding damage to early plant growth stages (Hentz & Nuessly, 2004), resulting in chlorosis and death of three pairs of leaves. Besides direct damage, yellow sugarcane aphid also transmits sugarcane mosaic potyvirus (Blackman & Eastop, 2000).

The *S. flava* was first recorded on *M. ×giganteus* in 2008 in leaf damage collected from seven locations of Indiana, Illinois, Nebraska, Kentucky, and Iowa (Bradshaw et al., 2010). Aphids were found on young to old plantings (1–21 year) in large populations. Aphids feeding leads to leaf death. The symptoms of *M. ×giganteus* infestations, specifically yellowing and redding of leaves, were very similar to sugarcane (Nuessly & Hentz, 2002) and sorghum (Costa-Arbulú et al., 2001). The *S. flava* damage has economic importance as a key factor of *M. ×giganteus* plantation establishment during the first year of vegetation.

Corn leaf aphid *Rhopalosiphum maidis* L. (Homoptera: Aphidiidae) is native to Asia, although they have spread almost worldwide. It occurs sporadically in cool temperature climate (Blackman & Eastop, 2000). In several countries it is considered as an economically important pest of *Poaceae* (*Gramineae*) monocot crops, namely, wheat, barley, sorghum, and *M. sinensis* (Carena & Glogoza, 2004; Huggett et al., 1999). This damage potential is especially concerning because most plant viruses are transmitted by aphids (Hull, 2002) and *R. maidis* can transmit the RPV strain (cereal yellow dwarf virus) of barley yellow dwarf luteovirus (BYDV) to *M. ×giganteus* (Huggett et al., 1999; Jarošová et al., 2013).

8.2.2.4 Identification

The wingless corn leaf aphid is oval, about 2 mm, blue-green in color with black antennae, legs, and tailpipes. The body and legs are black. The nymphs are similar to adults with smaller size and underdeveloped antennae and tailpipes. The winged form of the aphid is about the same size as the wingless ones, with dark green to black body and black tailpipes.

8.2.2.5 Life Cycle

The species reproduce via parthenogenesis. It overwinters in both eggs and females (in warmer climates) on cereal grasses. The optimum temperature for development is around 30°C. Corn leaf aphids live in large colonies on their host plants.

8.2.2.6 Damage

The corn leaf aphids feed by removing plant sap from the phloem. The typical signs of feeding while insect in high population are leaf mottling and/ or discoloration and reddening. The plant vulnerability to insect damage increases under drought stress. The aphids also produce honeydew, which covers the leaf surface. Mold fungi colonize the honeydew producing a black layer of fungal colonies on corn leaf surface causing the reduction of the photosynthetic leaf activity.

Natural *R. maidis* infestation of *M. ×giganteus* was found in Indiana, Illinois, Nebraska, Kentucky, and Iowa (USA) in 2008 survey. Species was observed on the young *M. ×giganteus* whorls in first-year plantings (Bradshaw et al., 2010). The infestation of young tillers resulted in yellowing of upper leaves and occurred later compared to *S. flava*.

Other aphids that specialize on *Poaceae* family that may infest Miscanthus in countries with temperate climates and vector of pathogens are bird-cherry oat aphid *Rhopalosiphum padi* L., wheat aphid *Schizaphis graminum* Rondani, and oat aphid *Sitobion avenae* Fabricius. Results of 2-years' field surveys conducted in 2010–2011 in three regions: Zhytomyr, Vinnitsa, and Kyiv in Ukraine (Stefanovska et al., 2017b) found that species with piercing-sucking moth part, namely, aphids and trips were most frequently observed among 50 herbivorous species which were recorded (Figure 8.1).

Bird-cherry oat aphid was recorded as a more frequently observed pest among *Aphidiidae* family on 1–5-year-old plantations in all three locations in this study with low population, attacking plants of first year stands, causing yellowing of leaves.

While carrying out the survey of *M. ×giganteus* in Lower Silesia (Poland), Hurej and Twardowski (2009) recorded *R. padi*, which was the dominant species among other aphids. The population was low, however, aphid attacked young plants of first year vegetation.

The field study of *M. ×sachariphlorus* herbivores in Northern France (Coulette et al., 2013) showed that population density of bird-cherry aphid *R. padi* which is generalist was the lowest among polyphagous species, i.e., green peach aphid *Myzus percicae* Sulzer and bean aphid *Aphis fabae* Scop. Results of laboratory experiment indicated that Miscanthus is not suitable for these three aphids and even it has the potential to act as a barrier to restrict spreading of phytoviruses (Coulette et al., 2013).

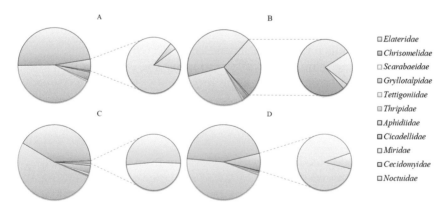

FIGURE 8.1
Frequency of occurrence of herbivorous individuals by families at the investigated locations in 2010–2011 growing seasons: A – Zhytomyr, B – Vinnitsa, C – Kyiv (1), D – Kyiv (2). (Modified from Stefanovska et al., 2017b.)

8.3 Plant Feeding Insects with Chewing Mouth Parts

8.3.1 Generalist Coleoptera

May beetle (*Melolontha melolontha* L.) and June beetle (*Amphimallon solstitialis* L.) (Coleoptera: Scarabaeidae) are the most destructive pests of turf grasses and cause economic threat as polyphagous pest of many fields, horticulture and orchard crops, including *Poaceae* (*Gramineae*) monocot crops which Miscanthus plants belong to.

8.3.1.1 Identification

Adults of May beetle are stocky insects that vary in color: shade of brown and tan to dark chocolate with length of 22–30 mm. The larvae (grubs) – fat, C-shaped whitish yellow grubs with brown heads in length of 40–45 mm in final instar.

8.3.1.2 Life Cycle

May beetles develop for 3–5 years. It hibernates in the stage of grubs and adults. Grubs live in the soil feeding on plant roots for several years. For pupating they move deeper in the soil, later emerging from the ground as adults in the spring.

8.3.1.3 Damage

Adult beetles eat up leaves and flowers, preferring trees. Grubs feed on dead organic matter and plant roots, resulting in visible water stress and ultimately ending in the death of the plant. The larval stages of these beetles are more harmful to crops.

M. *melolontha* was the dominant species observed (Stefanovska et al., 2017b) in two out of three regions in a field survey of *M.* ×*giganteus* growing in Ukraine (Figure 8.2). The reduction of *M.* ×*giganteus* seedlings at first year of vegetation due to feeding grubs on roots was observed in a field infected by *M. melolontha* L. and *A. solstitialis* L. in population density – 7.2 grubs m^{-2}. The second and third grub instar causes the biggest damage. Preplanting treatment of rhizomes by insecticides had positive effect on seedling emergence, survival, and further plant height (Sabluk et al., 2014).

8.3.2 Generalist Lepidoptera

Fall armyworm *Spodoptera frugiperda* Smith (Lepidoptera: Noctuidae) is one of the important polyphagous pests distributed in Brazil, Argentina, USA (Clark et al., 2007; Montezano et al., 2018), and Africa (De Groote et al., 2020).

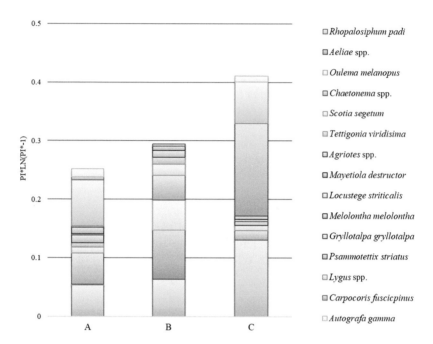

FIGURE 8.2
Variation of insect species and genera on *M. ×giganteus* in three location of Ukraine, 2010–2011.
A – Zhytomyr, B – Vinnitsa, C – Kyiv. (Modified from Stefanovska et al., 2017b.)

It caused significant yield reduction on corn, soybean, cotton, and diverse grasses; pest has two strains: rice and corn.

8.3.2.1 Identification

The moths have a wingspan of 32–40 mm. The adult moths have a brown or gray forewing and a white hindwing. Male fall armyworms have more patterns and a distinct white spot on each forewing. The hind wing is similar for both sexes: silver white with a narrow dark border. Larvae are a light green to dark brown with longitudinal stripes. In the sixth instar, larvae can reach 4.5 cm long (Brambila, 2009).

8.3.2.2 Life Cycle

The Fall armyworm hibernates in the stage of pupa; however, it is not tolerant to low temperature and can survive only in mild winters and when freezing occurs all stages are usually killed. Adults can migrate over 500 km. They lay eggs on the leaves of the host. Young larva hatch from eggs in 1–10 days and migrate to the whorl. The duration of life cycle depends on the season and in winter is longer-lasting. The number of generations is two (northern regions) and up to six (tropics).

8.3.2.3 Damage

Larvae cause damage on the leaves when they hatch by consuming foliage. Young larvae initially consume leaf tissue from one side, leaving the opposite epidermal layer intact. The older larvae begin to make holes in leaves and eat from the edge of the leaves inward. Feeding in the whorl of corn often produces a characteristic row of perforations in the leaves.

Fall armyworm *S. frugiperda* was found to infest the whorls of *M. ×giganteus* in field plots (Prasifka et al., 2009). Greenhouse study using fall armyworm population collected in the field from natural infestation indicated that both corn and rice strains of *S. frugiperda* are able to develop and reproduce on Miscanthus along with other hosts, including corn. In this experiment fall armyworm evidently preferred corn to Miscanthus; however, relative survival of rice strain fall armyworm was greater on Miscanthus. Thus, this dangerous pest can infest the biomass and food/feed crop (Prasifka et al., 2009).

Stefanovska et al. (2017b) observed two Noctuidae species in *M. ×giganteus*: Turnip Moth *Scotia segetum* L. and Silver Y-Moth *Autografa gamma* L., feeding on planting at first year of vegetation (Figure 8.2); however, no conspicuous symptoms were found.

8.3.3 Generalist Coleopteran

The Western corn rootworm *Diabrotica virgifera virgifera* Le Conte is a native to North America, where it is considered as the main pest of corn (Yu et al., 2019). The pest is present in many European countries, spreading constantly year by year (Toth et al., 2020). Corn is the only crop in which high population densities can develop.

8.3.3.1 Identification

The adults, 5–6 mm long, are dark yellow in color and there are three black stripes on the wing covers. Newly hatched larvae are nearly colorless, but gradually turn white as they feed and get older. Mature larvae reach about 13 mm long, and are creamy, white with a brown head capsule.

8.3.3.2 Life Cycle

Western corn rootworms produce one generation per year (Chiang, 1973). The pest overwinters in the soil in egg stage. The larvae hatch from eggs from late spring to early summer. The larvae move around in the soil, feeding on the maize roots. Their development lasts approximately 1 month. Larvae pupate in the soil and emerge as adults in 5–10 days. Adults emerge throughout the summer period and can mate several times. Females lay eggs in small clutches near the base of corn stalks, where they remain unhatched for the winter (Gray et al., 2009).

8.3.3.3 Damage

Pest causes damage in the stage of larvae who feed on both the leaves and the silk of the female inflorescence of corn. Larvae feeding reduced plants' ability to take water and nutrients. Initially, injured root tips are discolored or have brown lesions; after some time, primary or secondary roots can be completely pruned. Western corn rootworm produces a single generation each year (Yu et al., 2019).

Miscanthus is suitable for larval development for both European (Gloyna et al., 2011) and the US (Spencer et al., 2009) populations of the western corn rootworm. Since Western corn rootworm survives on *M. ×giganteus*, it has economic consequence. Without crop rotation and tillage, the risk of Western corn rootworm from *M. ×giganteus* as biomass crops into nearby food and feed crops, including corn may significantly increase. This is especially challenging in upscaling of *M. ×giganteus* production in agricultural and marginal land in intensive monoculture systems.

8.4 Plant Feeding Nematodes Associate with *M. × giganteus*

The limited number of reported nematodes in *M. ×giganteus* can be explained by restricted range of researches conducted in this area. Plant-feeding nematodes according to the feeding habits are commonly assigned to the following trophic groups: obligate plant feeders, facultative plant feeders that alternatively feed on fungi or bacteria, and fungal feeders that alternatively feed on plants (Yeates et al., 1993). Plant parasitic nematodes (PPNs) are a very important group of pests because of economical aspect of ability to decrease food and feed crops (Bernard et al., 2017). This nematode group is studied most profoundly (Emery et al., 2017). Because of the relatively small portion of *M. ×giganteus* plantations and short times since beginning of cultivation, this is still not clear to which extent PPNs can contribute to this crop yield loss.

In the frame of multistate parasitic nematode survey of 37 miscanthus and 48 switch grass plots across Iowa, Illinois, Georgia, Kentucky, South Dakota, and Tennessee (USA), the following PPNs were present: lesion (*Pratylenchus* spp.), needle (*Longidorus* spp.), dagger (*Xiphinema* spp.), lance (*Hoplolaimus* spp.), stunt (*Tylenchorynchus* spp.), spiral (*Helicotylenchus* spp.), and ring (*Criconema* spp.) (Mekete et al., 2009; Mekete et al., 2011a, b). The damage thresholds of *M. ×giganteus* PPNs have not been reported yet. Mekete et al. (2011b) used values existing for monocotyledon hosts to compare it with recorded PPNs densities. It was found that population densities of *Helicotylenchus*, *Xiphinema*, *Pratylenchus*, *Hoplolaimus*, *Tylenchorhynchus*, *Criconemella*, and *Longidorus* spp. exceeded threshold value ranges reported for another monocotyledon host.

8.4.1 PPNs – Potential Vector of Plant Viruses

The nematode survey associated with *M. ×giganteus* was provided in established crop's stands of 1–10 years of age representing, consequently, eight locations and six soil types in Ukraine (out of which three were contaminated by trace elements) and nine localities covering eight soil types in Poland (Figure 8.3).

The obtained results indicated that group of plant-feeding species was represented by 53 species belonging to 22 genera and 10 families (Stefanovska et al., 2020). Comparison of population density of plant-feeding nematodes recorded in both Ukraine and Poland with the damage threshold values established by Mekete et al. (2011b) demonstrated that the populations of several PPNs, which are known as vector of plant viruses, were above the estimated damage threshold.

Two dangerous plant viruses, specifically Norovirus (NEPO) and Tobravirus (TOBRA), have nematode vectors. The only known nematode vectors of NEPO are in the genera *Xiphinema* and *Longidorus* whereas TOBRA is transmitted by *Trichodorus* and *Paratrichodorus* (Macfarlane, 2003; MacFarlane & Neilson, 2009).

A total of ten nematode species capable of vectoring plant viruses were recorded in Ukraine and Poland in surveyed *M. × giganteus* plantations. Two species of the genus *Longidorus*: *L. elongatus* and *L. attenuatus* were observed only in Poland with a population range of 284–300 individuals per 100 cm^3. They exceeded the damage level (5–25 individuals per 100 cm^3) by 2.5-fold. Species of the genus *Xiphinema* were associated with *M. × giganteus* plantations in both countries with population densities below the estimating threshold. Nematodes of genera *Xiphinema* would not likely impact negatively the crop development and biomass yield in the observed plantations. These results are not inconsistent with Mekete et al. (2011b) who showed that this genus was potentially damaging for *M. × giganteus*. Emery et al. (2017) observed that Miscanthus is suitable to *Xiphinema*.

Three species of genus *Trichodorus* in Poland *T. sparsus*, *T. similis*, and *T. viruliferus* vs. one species in Ukraine *T. sparsus* were recorded. PPNs of genus *Paratrichodorus* recovered from the root zone of *M. ×giganteus* were represented by two species: *P. pachydermus* and *P. teres* in both countries.

8.4.2 Ecto-, Endoparasites, and Hyphal/Root Feeders

Nematode species representing ectoparasites belonging to the genera *Mesocriconema, Criconema, Paratylenchus, Geocenamus, Bitylenchus, Merlinius, Neodolichorhynchus, Sauertylenchus, Scutylenchus,* and *Amplimerlinius* were recovered at the maximum population densities of 60–70/100 cm^3 of soil. These values are considerably lower than in suggested damage threshold value that is estimated as 300–600 nematodes per 100 cm^3.

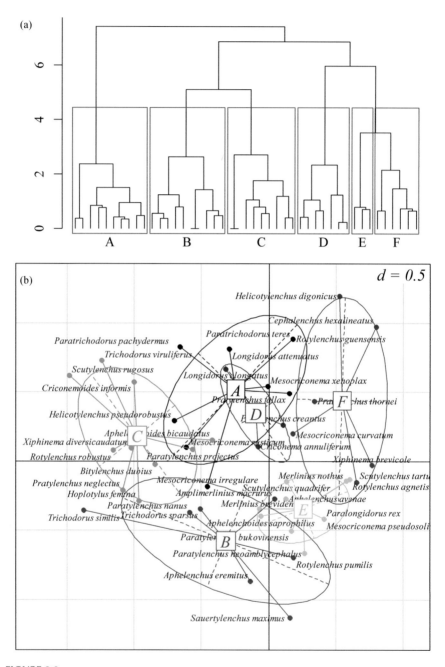

FIGURE 8.3
Cluster analysis and NMDS procedure, where (a) cluster arrangement, Y-scale on the chart is
the Euclidian distance between sampling points (presented solution with six clusters representing soil types A, B, C, D, E, and F); (b) the projection of clusters (A, B, C, D, E, and F) in the
space of the first two dimensions. (Modified from Stefanovska et al., 2020.)

Two families *Pratylenchus* and *Helicotylenchus* in survey samples represented endoparasite and semi-endoparasite nematodes. Four species of the lesion nematodes *Pratylenchus* were recovered in both countries. Population densities of *P. crenatus* and *P. neglectus* were greater than the estimated damage thresholds (50–100 individuals per 100 cm³ of soil) and can be viewed as the potential pathogen of *M. × giganteus*. *Pratylenchus* spp. are known as very dangerous plant pathogens responsible for lesion disease that rank third place in the world list of the most economically important species (Jones & Fosu-Nyarko, 2014). This finding for Ukraine and Poland was in agreement with results obtained earlier by Mekete et al. (2011a) in the US.

Spiral nematodes *Helicotylenchus* were found in densities of 300 individuals per 100 cm³, that is, lower densities than the established threshold value. Hyphal and root feeders were represented by nematodes from four genera: *Cylindrolaimus*, *Rhabditis*, *Plectus*, and *Anaplectus* that were recovered from samples with population densities up to 25 individuals per 100 cm³ of soil.

The characteristics of PPNs associated with *M. × giganteus* demonstrated (Mekete et al., 2011b) that economically important nematodes, specifically needle *Longidorus elongates* (*Longidoridae*) and root-lesion nematodes *Pratylenchus crenatus* and *Pratylechchus neglectus* (*Pratylenchidae*), share with *M. × giganteus* other hosts from the same plant family (*Gramineae*). The fact that all PPNs reported to feed on *M. × giganteus* are pests of corn, sugarcane, or sorghum raises concerns that the production of that crop in the large scale may increase pests not only on *M. × giganteus* but also on the numbers in existing food and feed crops.

8.4.3 The Indication of *M. × giganteus* Plantation State with Plant-Feeding Nematodes

M. × giganteus as a perennial grassland crop generally shows its maximum productivity in the third year of cultivation (Lewandowski et al., 2018). Knowledge based on interaction between PPNs and Miscanthus stands in different age is important to predict biomass productivity during the long-term growing. The PPNs community is determined both by temporal changes and environmental factors (Zhukov et al., 2018). Temporal stages of *M. × giganteus* plantations development in a wide geographical district by PPN species impact the plant damage under representative soil types.

The distribution of nematodes recorded in study in Ukraine and Poland was evaluated by applying the nonmetric multidimensional scaling approach (NMDS) indicated spatial heterogeneity of sampling points and community dynamics (Figure 8.3).

The relationships among the hyphal and root hair feeders, semi-endoparasites, and ectoparasites were suggested as indicators for assessing the state of *M. × giganteus* plantations with different years of cultivation with two aspects which were considered. It was determined (Figure 8.3) that the abundance of hyphal and root hair feeders expanded with increasing of plantation age, whereas the abundance of ectoparasites decreased. The increase in hyphal and root hair feeders with increasing age of planting was associated with the reductions in the number of all other trophic groups of parasitic nematodes. Therefore, the increase in hyphal and root hair feeders was the most important marker of the planting age of the *M. ×giganteus* (Stefanovska et al., 2020). This result was in a good agreement with findings of de Goede et al. (1993) which showed an increase in the number of hyphal and root hair feeders in the successional series of crop development.

The research studies carried out for 20 years indicate that intensive involvement of *M. ×giganteus* into biomass systems in conventional agriculture and phytotechnologies stimulates several problems, specifically, worsening the pest problem at other hosts from the same plant family and direct damage of Miscanthus that negatively impacts crop yield. However, considering the perennial nature of *M. ×giganteus*, its long cultivation may support the increase of biocontrol agents, capable to regulate pest population and keep it under damaging level. It is still not clear to what extent these groups of organisms contribute to yield reduction, and consequently needs for pest management systems that should have been addressed are rather uncertain.

References

Ben Fradj, N., Rozakis, S., Borzęcka, M., & Matyka, M. (2020). Miscanthus in the European bio-economy: A network analysis. *Industrial Crops and Products, 148*, 112281. https://doi.org/10.1016/j.indcrop.2020.112281.

Bernard, G. C., Egnin, M., & Bonsi, C. (2017). The impact of plant-parasitic nematodes on agriculture and methods of control. In: Shah, M. M., & Mahamood, M., eds. *Nematology Concepts, Characteristics and Control.* InTech, London. https://doi.org/10.5772/intechopen.68958.

Blackman, R. L., & Eastop, V. F. (2000). *Aphids on the World's Crop: An Identification Guide* (2nd ed.). John Wiley & Sons Ltd, 476. ISBN: 978-0-471-85191-2.

Bradshaw, J. D., Prasifka, J. R., Steffey, K. L., & Gray, M. E. (2010). First report of field populations of two potential aphid pests of the bioenergy crop *Miscanthus ×giganteus. The Florida Entomologist, 93*(1), 135–137. https://doi.org/10.1653/024.093.0123.

Brambila, J. (2009). Steps for the dissection of male Spodoptera moths (Lepidoptera: Noctuidae) and notes on distinguishing S. litura and S. littoralis from native Spodoptera species. USDA-APHIS-PPQ.

Carena, M. J., & Glogoza, P. (2004). Resistance of maize to the corn leaf aphid: a review. *Maydica, 49*(4), 241–254. https://doi.org/10.1007/s10681-009-0044-z.

Chiang, H. C. (1973). Bionomics of the Northern and Western corn rootworms. *Annual Review of Entomology, 18*(1), 47–72. https://doi.org/10.1146/annurev.en.18.010173.000403.

Clark, P. L., Molina-Ochoa, J., Martinelli, S., Skoda, S. R., Isenhour, D. J., Lee, D. J., Krumm, J. T., & Foster, J. E. (2007). Population variation of the fall armyworm, *Spodoptera frugiperda*, in the Western Hemisphere. *Journal of Insect Science, 7*(5), 1–10. https://doi.org/10.1673/031.007.0501.

Cook, J. H., Beyea, J., & Keeler, K. H. (1991). Potential impacts of biomass production in the United States on biological diversity. *Annual Review of Energy and the Environment, 16*(1), 401–431.

Costa-Arbulú, C., Gianoli, E., Gonzáles, W. L., & Niemeyer, H. M. (2001). Feeding by the aphid *Sipha flava* produces a reddish spot on leaves of *Sorghum halepense*: An induced defense? *Journal of Chemical Ecology, 27*(2), 273–283. https://doi.org/10.1023/A:1005676321251.

Coulette, Q., Couty, A., Lasue, P., Rambaud, C., & Ameline, A. (2013). Colonization of the biomass energy crop Miscanthus by the three aphid species, *Aphis fabae, Myzus persicae*, and *Rhopalosiphum padi*. *Journal of Economic Entomology, 106*(2), 683–689. https://doi.org/10.1603/EC12147.

de Goede, R. G. M., Verschoor, B. C., & GeocevA, S. S. (1993). Nematode distribution, trophic structure and biomass in a primary succession of blown-out areas. *Fundamental and Applied Nematology, 16*(6), 525–538.

De Groote, H., Kimenju, S. C., Munyua, B., Palmas, S., Kassie, M., & Bruce, A. (2020). Spread and impact of fall armyworm (*Spodoptera frugiperda* J.E. Smith) in maize production areas of Kenya. *Agriculture, Ecosystems and Environment, 292*, 106804. https://doi.org/10.1016/j.agee.2019.106804.

Elshout, P. M. F., Zelm, R., Velde, M., Steinmann, Z., & Huijbregts, M. A. J. (2019). Global relative species loss due to first-generation biofuel production for the transport sector. *GCB Bioenergy, 11*(6), 763–772. https://doi.org/10.1111/gcbb.12597.

Emery, S. M., Reid, M. L, Bell-Dereske, L, & Gross, K. L. (2017). Soil mycorrhizal and nematode diversity vary in response to bioenergy crop identity and fertilization. *GCB Bioenergy, 9*(11), 1644–1656. https://doi.org/10.1111/gcbb.12460.

Gloyna, K., Thieme, T., & Zellner, M. (2011). Miscanthus, a host for larvae of a European population of Diabrotica v. virgifera. *Journal of Applied Entomology, 135*(10), 780–785. https://doi.org/10.1111/j.1439-0418.2010.01599.x.

Gołąb-Bogacz, I., Helios, W., Kotecki, A., Kozak, M., & Jama-Rodzeńska, A. (2020). The influence of three years of supplemental nitrogen on above- and belowground biomass partitioning in a decade-old *Miscanthus × giganteus* in the Lower Silesian Voivodeship (Poland). *Agriculture, 10*(10), 473. https://doi.org/10.3390/agriculture10100473.

Gray, M. E., Sappington, T. W., Miller, N. J., Moeser, J., & Bohn, M. O. (2009). Adaptation and invasiveness of Western corn rootworm: Intensifying research on a worsening pest. *Annual Review of Entomology, 54*(1), 303–321. https://doi.org/10.1146/annurev.ento.54.110807.090434.

Gruss, I., Stefanovska, T., Twardowski, J., Pidlisnyuk, V., & Shapoval, P. (2019). The ecological risk assessment of soil contamination with Ti and Fe at military sites in Ukraine: Avoidance and reproduction tests with *Folsomia candida*. *Reviews on Environmental Health, 34*(3), 303–307. https://doi.org/10.1515/reveh-2018-0067.

Hentz, M., & Nuessly, G. (2002). Morphology and biology of *Diomus terminatus* (Coleoptera: Coccinellidae), a predator of *Sipha flava* (Homoptera: Aphididae). *Florida Enthomologist, 85*(1), 276–279. https://doi.org/10.1653/0015-4040(2002)085[0276:mabodt]2.0.co;2.

Hentz, M., & Nuessly, G. (2004). Development, longevity, and fecundity of *Sipha flava* (Homoptera: Aphididae) feeding on *Sorghum bicolor*. *Environmental Entomology, 33*(3), 546–553. https://doi.org/10.1603/0046-225X-33.3.546.

Huggett, D. A. J., Leather, S. R., & Walters, K. F. A. (1999). Suitability of the biomass crop *Miscanthus sinensis* as a host for the aphids *Rhopalosiphum padi* (L.) and *Rhopalosiphum maidis* (F.), and its susceptibility to the plant luteovirus Barley Yellow Dwarf Virus. *Agricultural and Forest Entomology*, *1*(2), 143–149. https://doi.org/10.1046/j.1461-9563.1999.00019.x.

Hull, R. (2002). *Matthews' Plant Virology*. Academic Press, London.

Hurej, M., & Twardowski, J. (2009). Phytophagous insects on *Miscanthus giganteus* (*Miscanthus × giganteus* L.). *Progress in Plant Protection*, *49*(3), 1183–1186.

Jarošová, J., Chrpová, J., Šíp, V., & Kundu, J. K. (2013). A comparative study of the Barley yellow dwarf virus species PAV and PAS: distribution, accumulation and host resistance. *Plant Pathology*, *62*(2), 436–443. https://doi.org/10.1111/j.1365-3059.2012.02644.x.

Jones, M. G. K., & Fosu-Nyarko, J. (2014). Molecular biology of root lesion nematodes (*Pratylenchus spp.*) and their interaction with host plants. *Annals of Applied Biology*, *164*(2), 163–181. https://doi.org/10.1111/aab.12105.

Kindler, S. D., & Dalrymple, R. L. (1999). Relative susceptibility of cereals and pasture grasses to the yellow sugarcane aphid (*Homoptera: Aphididae*). *Journal of Agricultural and Urban Entomology*, *16*(2), 113–122.

Kosztarab, M., & Kozár, F. (2012). *Scale Insects of Central Europe* (Vol. 40). Springer Science & Business Media, Dordrecht, Netherlands.

Landis, D. A., & Werling, B. P. (2010). Arthropods and biofuel production systems in North America. *Insect Science*, *17*(3), 220–236. https://doi.org/10.1111/j.1744-7917.2009.01310.x.

Lewandowski, I., Clifton-Brown, J., Kiesel, A., Hastings, A., & Iqbal, Y. (2018). Miscanthus. In Alexopoulou, E., ed. *Perennial Grasses for Bioenergy and Bioproducts: Production, Uses, Sustainability and Markets for Giant Reed, Miscanthus, Switchgrass, Reed Canary Grass and Bamboo*. Academic Press, London (pp. 35–55). https://doi.org/10.1016/B978-0-12-812900-5.00002-3.

Macfarlane, S. A. (2003). Molecular determinants of the transmission of plant viruses by nematodes. *Molecular Plant Pathology*, *4*(3), 211–215. https://doi.org/10.1046/j.1364-3703.2003.00164.x.

MacFarlane, S. A., & Neilson, R. (2009). Testing of transmission of tobraviruses by nematodes. *Current Protocols in Microbiology*, *12*(1), 16B.5.1–16B.5.16. https://doi.org/10.1002/9780471729259.mc16b05s12.

Mekete, T., Gray, M. E., & Niblack, T. L. (2009). Distribution, morphological description, and molecular characterization of *Xiphinema* and *Longidorous spp.* associated with plants, *Miscanthus spp.* and *Panicum virgatum* used for biofuels. *Global Change Biology Bioenergy*, *1*, 257–266. https://doi.org/10.1111/j.1757-1707.2009.01020.x.

Mekete, T., Reynolds, K., Lopez-Nicora, H. D., Gray, M. E., & Niblack, T. L. (2011a). Distribution and diversity of root-lession nematode (*Pratylenchus spp.*) associated with Miscanthus × *giganteus* and *Panicum virgatum* used for biofuels, and species identification in a multiplex polymerase chain reaction. *Nematology*, *13*(6), 673–686. https://doi.org/10.1163/138855410X538153.

Mekete, T., Reynolds, K., Lopez-Nicora, H. D., Gray, M. E., & Niblack, T. L. (2011b). Plant-parasitic nematodes are potential pathogens of Miscanthus × *giganteus* and *Panicum virgatum* used for biofuels. *Plant Disease*, *95*(4), 413–418. https://doi.org/10.1094/PDIS-05-10-0335.

Montezano, D. G., Specht, A., Sosa-Gómez, D. R., Roque-Specht, V. F., Sousa-Silva, J. C., Paula-Moraes, S. V., Peterson, J. A., & Hunt, T. E. (2018). Host plants of *Spodoptera frugiperda* (*Lepidoptera: Noctuidae*) in the Americas. *African Entomology, 26*(2), 286–300. https://doi.org/10.4001/003.026.0286.

Nuessly, G., & Hentz, M. (2002). Feeding effects of yellow sugarcane aphid on sugarcane. *Journal of the American Society of Sugar Cane Technologists, 22,* 126–127.

Pallipparambil, G. R., Cha, G., & Gray, M. E. (2014). A comparative life-table analysis of *Sipha flava* (*Hemiptera: Aphididae*) on two biofuel hosts, *Miscanthus × giganteus* and *Saccharum spp. Journal of Economic Entomology, 107*(3), 1069–1075. https://doi.org/10.1603/EC13263.

Prasifka, J. R., Bradshaw, J. D., Meagher, R. L., Nagoshi, R. N., Steffey, K. L., & Gray, M. E. (2009). Development and feeding of fall armyworm on *Miscanthus × giganteus* and switchgrass. *Journal of Economic Entomology, 102*(6), 2154-2159. https://doi.org/10.1603/029.102.0619.

Reid, W. V., Ali, M. K., & Field, C. B. (2020). The future of bioenergy. *Global Change Biology, 26*(1), 274–286. https://doi.org/10.1111/gcb.14883.

Sabluk, V. T., Gryschenko, O. M., & Stefanovska, T. R. (2014). Control of grub population in plantations of energy willow and giant miscanthus (*Miscanthus × giganteus*). *Bioenergetyka (Bioenergy), 2,* 31–32.

Semere, T., & Slater, F. M. (2007). Invertebrate populations in miscanthus (*Miscanthus × giganteus*) and reed canary-grass (*Phalaris arundinacea*) fields. *Biomass and Bioenergy, 31*(1), 30–39. https://doi.org/10.1016/j.biombioe.2006.07.002.

Spencer, J. L., Hibbard, B. E., Moeser, J., & Onstad, D. W. (2009). Behaviour and ecology of the western corn rootworm (*Diabrotica virgifera virgifera* LeConte). *Agricultural and Forest Entomology, 11*(1), 9–27. https://doi.org/10.1111/j.1461-9563.2008.00399.x.

Stefanovska, T., Chumak, P., Pidlisnyuk, V., Kovalchuk, V., Kava, L., & Likar, Ya. (2017a). State of the art on adaptability of *Miscanthiococcus miscanthi* Takaschi (*Homoptera, Pseudococcidae*) followed by the introduction of *Miscanthus × giganteus* to Ukraine. *Communications in Applied and Biological Sciences, 82*(2), 177–181. ISSN 1379-1176.

Stefanovska, T., Lewis, E., Pidlisnyuk, V., & Smyrnykh, O. (2015). First record of *Clytra laeviuscula* Ratzeburg as potential insect pest of energy willow (*Salix viminalis* L.) in Ukraine. *Agriculture (Pol'nohospodárstvo), 61*(3), 115–118. https://doi.org/10.1515/agri-2015-0016.

Stefanovska, T., Pidlisnyuk, V., Lewis, E., & Gorbatenko, A. (2017b). Herbivorous insects diversity at *Miscanthus × giganteus* in Ukraine. *Agriculture, 63*(1), 23–32. https://doi.org/10.1515/agri-2017-0003.

Stefanovska, T., Skwiercz, A., Zouhar, M., Pidlisnyuk, V., & Zhukov, O. (2020). Plant-feeding nematodes associated with *Miscanthus × giganteus* and their use as potential indicators of the plantations' state. *International Journal of Environmental Science and Technology,* 1–16. https://doi.org/10.1007/s13762-020-02865-z.

Thomson, L. J., & Hoffmann, A. A. (2011). Pest management challenges for biofuel crop production. *Current Opinion in Environmental Sustainability, 3*(1–2), 95–99. https://doi.org/10.1016/j.cosust.2010.11.003.

Toth, S., Szalai, M., Kiss, J., & Toepfer, S. (2020). Missing temporal effects of soil insecticides and entomopathogenic nematodes in reducing the maize pest *Diabrotica virgifera virgifera. Journal of Pest Science, 93*(2), 767–781. https://doi.org/10.1007/s10340-019-01185-7.

Warner, E., Inman, D., Kunstman, B., Bush, B., Vimmerstedt, L., Peterson, S., Macknick, J., & Zhang, Y. (2013). Modeling biofuel expansion effects on land use change dynamics. *Environmental Research Letters, 8*(1), 015003. https://doi.org/10.1088/1748-9326/8/1/015003.

Werling, B. P., Meehan, T. D., Robertson, B. A., Gratton, C., & Landis, D. A. (2011). Biocontrol potential varies with changes in biofuel-crop plant communities and landscape perenniality. *GCB Bioenergy, 3*(5), 347–359. https://doi.org/10.1111/j.1757-1707.2011.01092.x.

Wheeler, A. G. (2013). *Hyperaspis paludicola* Schwarz (*Coleoptera: Coccinellidae*): Association with Miscanthus mealybug, *Miscanthicoccus miscanthi* (Takahashi) (*Hemiptera: Pseudococcidae*), with a review of its distribution. *Entomological News, 123*(3), 220–224. https://doi.org/10.3157/021.123.0308.

Winkler, B., Mangold, A., von Cossel, M., Clifton-Brown, J., Pogrzeba, M., Lewandowski, I., Iqbal, Y., & Kiesel, A. (2020). Implementing miscanthus into farming systems: A review of agronomic practices, capital and labour demand. *Renewable and Sustainable Energy Reviews, 132*, 110053. https://doi.org/10.1016/j.rser.2020.110053.

Yeates, G. W., Bongers, T., De Goede, R. G., Freckman, D. W., & Georgieva, S. S. (1993). Feeding habits in soil nematode families and genera-an outline for soil ecologists. *Journal of Nematology, 25*, 315–331.

Yu, E. Y., Gassmann, A. J., & Sappington, T. W. (2019). Effects of larval density on dispersal and fecundity of western corn rootworm, *Diabrotica virgifera virgifera* LeConte (*Coleoptera: Chrysomelidae*). *PLoS One, 14*(3), e0212696. https://doi.org/10.1371/journal.pone.0212696.

Zhukov, O. V., Kunah, O. M., Dubinina, Y. Y., & Novikova, V. O. (2018). The role of edaphic, vegetational and spatial factors in structuring soil animal communities in a floodplain forest of the Dnipro river. *Folia Oecologica, 45*(1), 8–23. https://doi.org/10.2478/foecol-2018-0002.

9

Economics of Phytoremediation with Biomass Production

Larry E. Erickson, Jan Černý, and Valentina Pidlisnyuk

Abstract

The economics associated with phytotechnologies includes environmental, social, and ecosystem costs and benefits associated with the project. There are local costs and benefits associated with the site as well as global benefits because of carbon sequestration. Improvements in soil quality have long-term benefits and increase the value of the land. Risk reduction has health and safety benefits as well as improved value for the land. For many contaminated sites, there are many benefits associated with phytoremediation with biomass production that have value for society such as improved aesthetic values, better conditions for wildlife, employment benefits because of the project, and better quality of life in the community. While it would be great if the biomass produced would fully cover project costs, this should not be expected. One of the most important benefits associated with phytotechnologies is the addition of soil organic matter and the associated improvements in the health of the biological populations that are beneficial to plant productivity. The economics of phytoremediation with biomass production using Miscanthus is included in this chapter.

CONTENTS

9.1 Introduction to Phytoremediation with Biomass Production

In this chapter, the economic aspects of phytoremediation with biomass production are addressed with full consideration of environmental, social, ecosystem, and economic benefits associated with the improvement of each contaminated site using a sustainable remediation approach. The benefits of soil remediation with biomass production include risk reduction, improvements in soil quality and soil health, biomass products, carbon sequestration, reduced soil erosion, community aesthetic benefits, and better habitat for birds and wild animals.

9.2 Sustainable Approach

When economic considerations are integrated with environmental impacts and social values, environmental metrics such as air quality, water quality, soil health, and ecosystem conditions are introduced. Social metrics of human health, quality of life, safety, aesthetic value, and impact on quality of employment can be considered. The economic analysis should include both direct and indirect costs and benefits.

The economics of greenhouse gas (GHG) emissions based on the avoided social cost of carbon is part of the sustainable approach. Increasing the amount of soil organic matter and soil organic carbon is beneficial in establishing vegetation, and this can be considered in the global carbon balance. In the majority of phytoremediation with biomass production projects, soil organic carbon will be increased and biomass will be produced. Policies that provide incentives to reduce GHG emissions or increase soil organic carbon may be included in the analysis (Mikhailova et al., 2019). The Paris agreement on climate change and those working to reduce GHG emissions have included initiatives to increase soil carbon as part of the Paris climate pledges (Paustian et al., 2019; Rumpel et al., 2018). The National Academies have described a new research agenda on carbon sequestration in soils to help achieve the goals of the Paris agreement (NASEM, 2019).

There are many sustainability indicators that have been proposed for use in a sustainable remediation framework (Bardos et al., 2018). The International Sustainable Remediation Alliance has been established to encourage networking among organizations and countries that are making use of sustainable approaches to remediation of contaminated sites (Bardos et al., 2018). One of the results of using sustainability indicators has been to find that qualitative methods often lead to simple sustainability assessments that produce good decisions that are supported by the participants. When working with community representatives, there is value in using methods and concepts that are understood by all participants, especially when this leads to a good remediation plan.

The System of Environmental Economic Accounting (SEEA) has been developed to include the value of ecosystems in evaluating contaminated sites; the system includes the ability to include soil quality and carbon sequestration in ecosystem accounting (Hein et al., 2020; SEEA, 2014). Natural capital has value for society, and there is a global effort to extend accounting to include the value of ecosystems so that site remediation benefits can be fully communicated.

9.3 Benefits of Remediation

The total amount of land with contaminated soil is great. Values in the literature include that 28.3% of the land area in Europe has soil contamination (Ben Fradj et al., 2020), and more than 20 million ha worldwide are contaminated (Evangelou et al., 2012). The estimated amount in the United States is about 9 million ha (U.S. EPA, 2011). In addition, there are between 1 and 6 billion ha of degraded soil that need to have soil organic matter added to improve soil quality and land productivity (Rumpel et al., 2018).

As population increases it becomes more important to remediate contaminated sites and use the land for beneficial purposes. There are many benefits associated with converting an unused contaminated site into a site that is used for biomass production. By addressing the contamination such that risk is reduced, the site can be used with vegetation to consume carbon dioxide and produce oxygen and plant biomass that can be used by society. Soil quality and soil health can be improved by adding soil amendments such as manure and/or biosolids from wastewater treatment to increase soil organic carbon.

There is value to establishing Miscanthus or another biomass crop on the site that produces a useful product and adds soil carbon to the root zone. Soil organic carbon improves nutrient cycling, the ecological functioning of the organisms in the root zone soil, and the water holding capacity of the soil (Lal, 2016). Better soil health is one of the major ecosystem benefits of phytoremediation with biomass production.

The increase in soil carbon in the root zone has value with respect to carbon sequestration for the global carbon balance and Paris agreement on climate change. At sites with metals in the soil, adding organic carbon to soil reduces the availability of the metals.

The benefits of phytoremediation include the aesthetic value of a green site that provides better habitat for native wildlife compared to the site prior to remediation. Ornamental vegetation can be used for phytoremediation where parks or arboretums are established for public use and enjoyment

(Capuana, 2020). Phytoremediation reduces soil erosion and the spread of contaminants because of wind and water.

The social value of phytoremediation with biomass includes employment to carry out the remediation and to harvest the biomass. There would also be employment associated with the conversion of the biomass into products. The quality of life and safety near the site should be improved.

The economic costs and benefits include the costs associated with the phytoremediation with biomass production project. There are costs for site characterization, development of the remediation plan, and implementation of the plan. There are annual costs of caring for the site, harvesting the biomass, and selling the product. The annual income includes the net receipts for the biomass. There is economic value associated with the improved soil quality and the carbon sequestration, also.

The estimated value of the avoided social cost of carbon is $42.00/metric ton of carbon dioxide (Mikhailova et al., 2019; U.S. EPA, 2016). This is the estimated global value of reducing GHG emissions in the form of carbon dioxide by 1 metric ton. This is an appropriate value to use when the social benefit of adding soil organic carbon to a remediation project is to be included in the analysis.

The concept of reducing GHG emissions by adding organic carbon to soil has received significant attention, and the idea has strong support (Lal, 2016; Mikhailova et al., 2019; NASEM, 2019).

There is value in improving soil quality at contaminated sites. Many of these sites have little or no economic value in their present state; however, if the soil can be improved by establishing vegetation and producing a biomass product such as Miscanthus or wood that can be harvested and sold, the value of the site increases. Since productive land sells for more than $10,000/ha in many locations, one of the ecosystem benefits of phytoremediation with biomass production is the improvement in soil quality associated with the project. If one can remediate a contaminated site so that it can be used productively with products of the same quality as nearby farms, this is beneficial for society.

One way to improve the global carbon balance is to use phytoremediation with biomass production to establish trees on contaminated sites. Growing trees will increase soil carbon below ground and plant carbon above ground. At $42.00/ton of carbon dioxide avoided, it may be beneficial to do this at many sites. The soil carbon below Miscanthus after 8–10 years is about 90 MgC ha^{-1} (Cattaneo et al., 2014; Dondini et al., 2009; Hansen et al., 2004), which adds about $13,860 ha to the value of the project. Trees also have significant amounts of carbon in the soil and associated with their roots (Mikhailova et al., 2019). The amount of carbon added assumes that the site has very low soil carbon prior to the start of the remediation project.

The improvement in soil quality and the carbon added to the site are important ecosystem and economic reasons to go forward with phytoremediation projects. If one can move from the state of an abandoned site to one where a product is being produced, this has social value because of the employment that is created.

The cost of reforestation to reduce the amount of carbon dioxide in the atmosphere of \$1–\$10/ton of carbon dioxide is relatively low compared to many alternatives (Gillingham & Stock, 2018). If this is accomplished using phytoremediation at a contaminated site, land is restored to a productive use. In their work, they show that adding carbon to soil and aboveground vegetation is inexpensive compared to capturing carbon dioxide from combustion processes.

9.4 Motivation for Action

Vegetation is established at sites for several different reasons. Regulatory agencies may require land owners to reduce risk and/or contaminant movement due to wind and/or water erosion. Land owners may want to improve soil quality in order to use the land productively. There may be an interest to improve aesthetics and the quality of life in the neighborhood. A forest may be established to grow trees that benefit the global carbon balance and accumulate wood for future use.

The Tri-State Mining area in southeast Kansas, northeast Oklahoma, and southwest Missouri includes more than 80,000 ha (800 km^2). While there has been some remediation, there is a need to do more phytostabilization to reduce risk and improve soil quality so the land can be used more productively. After the mining ended, there has been some research related to establishing vegetation on the mine tailings that are found in southeast Kansas. This abandoned mine land is a good example of a site where organic matter such as manure or biosolids need to be added to help establish vegetation. The motivation at this site includes risk reduction, carbon sequestration, better aesthetics, soil quality improvement, and future profitable use of the land.

There is a significant effort to meet the goals of the Paris agreement on climate change. Each contaminated site that is restored using phytoremediation with added soil carbon and productive vegetation contributes positively to the effort to stop the accumulation of carbon in the atmosphere. There are many good choices with respect to what to plant. Locally adapted vegetation should be considered. It is also important to improve the ecosystem at the site so more and better ecosystem services are able to be provided.

9.5 Economics of Phytoremediation

The global interest and significant use of phytoremediation is because of its low cost compared to other methods (Pivetz, 2001). There have been many sites where this technology has been used successfully. The cost of using vegetation is often less than half of that for alternative technologies (Fiorenza et al., 1999; McCutcheon & Schnoor, 2004; Pivetz, 2001).

Phytoremediation has received much attention because the ecosystem at the contaminated site has generally been improved at sites where it has been implemented. The costs associated with growing vegetation are modest and supporting equipment is available because agriculture is global in its reach. Many people know how to grow plants and harvest produce.

The major costs associated with phytoremediation at field sites have often been associated with establishment of vegetation because amendments are needed to enable plants to grow at the site. The cost of amendments, moving them to the site, and incorporating them into the soil depend on the amount needed, distance transported, and process of incorporation into the soil at the site.

Analytical laboratory expenses can be one of the significant costs associated with phytoremediation. If the site is used for production of Miscanthus or to grow trees, the need for chemical analysis is reduced compared to growing food crops on the site. The analytical expense to characterize the contamination at the site is independent of the remediation method and often completed before a remediation method is selected.

9.6 Economics of Biomass Production

One of the goals of this book and the authors is to reduce the cost of remediation by producing products on contaminated sites that can be harvested and marketed. Miscanthus has significant biomass production and many potential uses. There has been great progress in developing Miscanthus as a crop to produce on contaminated sites. The economics of growing Miscanthus have been described in several publications (Hastings, 2017; Khanna et al., 2008; Witzel & Finger, 2016). The cost of establishment of Miscanthus using rhizomes is expensive, and there is great interest in developing less expensive methods such as planting seeds (Ben Fradj et al., 2020; Hastings, 2017). Witzel and Finger (2016) have reviewed 51 economic studies of Miscanthus in Europe and North America. They point out that location and regional demand for the Miscanthus impact the economics. Biomass yields vary with soil quality and weather, and these factors affect the economics.

Miscanthus as harvested has a relatively low density, and this impacts transportation costs and the cost of delivery to markets. If it is used for bedding or as a solid fuel near to where it is produced, transportation costs are small. It can be pelleted to increase the density and make it easier to handle and transport.

Miscanthus has high-quality cellulose which has value in making some products; however, there are many sources of cellulose including trees, wheat straw, and switchgrass. Thus, the price for Miscanthus is related to the general price for biomass that contains cellulose. Paper products can be made from Miscanthus. Large quantities of biomass are used in many parts of the world for making many types of paper.

As greater efforts are made to reduce GHG emissions, more policies to encourage renewable sources of energy may be approved. If the combustion of coal would have a charge of $42/Mg of carbon dioxide added to it for each *Mg = mega gram* produced, this would make using Miscanthus biomass in a coal-fired power plant much more competitive because the cost of using coal would more than double in the USA.

Miscanthus and other biomass crops that may be produced using phytoremediation at a contaminated site can be used as feed to an anaerobic digester or as feed to an ethanol plant. The price of natural gas would be increased by about $2.19/million BTUs if the $42/Mg carbon dioxide charge would be added. This value is $2.08/million kJ using metric units. Adding this value to the price of natural gas makes methane from renewable sources much more competitive.

One approach to including the social value of carbon in decision-making would be to have a global carbon tax of $42/Mg carbon dioxide on mined fuels with the tax revenue used to support projects that add soil carbon to improve agriculture and restore contaminated sites using phytoremediation with biomass production.

9.7 Bioeconomy of Miscanthus in Europe

The modern bioenergy provided 5.1% of total final energy demand in 2018, accounting for around half of all renewable energy in final energy consumption. Bioenergy provides around 9% of industrial heat demand and is concentrated in bio-based industries such as paper and board. Biofuels, mostly ethanol and biodiesel, provide around 3% of transport energy, and global biofuels production increased 5% in 2019 (REN21, 2020).

The European Union (EU) is the global leader in modern bioenergy production. In 2011, EU reaffirmed its objective to reduce GHG emissions by 80%–95% by 2050 compared to 1990 levels (Communication from the Commission to the European Parliament, 2011; Zappa et al., 2019). The Europe 2020 (2010) strategy seeks to address structural weaknesses in the economies of EU individual member states, economic and social problems, and pressure to reduce the proportion of nonrenewable (fossil) fuels and replace them with renewable energy sources (RESs) in the overall energy mix (solar, wind, geothermal, hydropower, and biomass energy) which account about 14% of global energy production. In between 2000 and 2015 EU more than doubled the share of bioenergy in gross final energy consumption (Tsemekidi Tzeiranaki et al., 2020). EU 2030 strategy (2017) called for sharing of renewable energy in gross energy consumption for at least 32% by 2030 with annual growth of RES min 1.3%, which will allow to go beyond EU commitment under the Paris Climate Agreement to reduce GHG emissions by at least 40% by 2030, compared to 1990 level (European Commission, 2019; RER 2030, 2019).

Germany is the largest producer of biomass which forms 23.6% of renewable electricity, and this share increased by 82.8% from 2008 to 2018 (Winkler et al., 2020). Providing sustainably produced feedstock for a growing bioeconomy is an important contribution to increase decarbonization of the German economy (Kiesel, 2020). In 2017 in the Czech Republic gross production of electricity from renewable sources accounted for 11.1% of the total domestic gross electricity production, and the share of renewable energy in primary energy sources was approximately 10.5% in 2017 (Ministry of Industry and Trade of the Czech Republic, 2018).

Perennial crops appear to be ideal for EU bioeconomy development; however, their cultivation plays a minor role in EU agriculture, and only about 43,800 ha of agricultural land were used for their production in 2015 (Cosentino et al., 2018). It may be explained by uncertainties about the economic viability and financial returns of these relatively novel crops, the long-term allocation of agricultural land to their production, the high investment costs for initial establishment (Sherrington et al., 2008), and absent of markets for biomass from perennial crops, in particular, from Miscanthus (Lewandowski et al., 2016; Witzel & Finger, 2016).

However, since 2018, Miscanthus has been included in the so-called "Greening" of the EU (Regulation (EU) 2017/2393), which might be advantageous for farmers to cultivate Miscanthus. In Central and Eastern Europe Miscanthus has a potential for bioenergy and bioeconomy (Dubis et al., 2019; Kvak et al., 2018; Tryboi, 2018). Biomass of this crop is becoming among important sources of energy in Ukraine (Roik et al., 2019) as country suffers from lack of imported fossil fuels and looks for substitution sources (Geletukha, 2017). By 2035 the share of renewable energy in the Ukrainian energy balance is expected to be 25%, with essential input of bioenergy plants. Two types of energy crops are cultivated in Ukraine currently, fast-growing willow (79%) and *M. × giganteus* (15%) (Geletukha et al., 2016).

The comprehensive literature review (Witzel & Finger, 2016) overviewed 51 scientific papers dealing with economics of Miscanthus cultivation to reveal the factors influencing the adoption decision of farmers. The most crucial factors for the profitability of Miscanthus and adoption decision of the farmer are diverse, including the following aspects: the expected lifespan, biomass yields, prices, establishment and opportunity costs, and subsidization possibilities. The large uncertainty is concerning the key parameters: yields and prices. Across 51 reviewed studies, mean yield assumptions ranged from 10 to 48 t dry biomass ha^{-1} while the assumptions concerning mean prices range from €48 to €134/ton of dry mass. And absence of an established market for biomass is mentioned as a major impediment leading to increased market risks for Miscanthus feedstock.

The perspective of bioenergy crop cultivation has to fulfill the following requirements (Von Cossel et al., 2019):

- providing a beneficial social and ecological contribution in increasing of agro-ecological biodiversity and landscape aesthetics;
- ensuring cultivation on marginal agricultural land or slightly contaminated land which allow to avoid competition with food crop production; bioenergy crops have to be able to cope with the given biophysical constraints on marginal agricultural lands;
- resilience of bioenergy cropping systems toward the growing climate change effect;
- fostering rural development and supporting the vast number of small-scale family farmers, managing some 80% of the global agricultural land and natural resources.

The production costs and labor requirement of Miscanthus cultivation and processing to bioproducts: animal bedding, combustion, and biogas were assessed recently (Winkler et al., 2020). The approach based on summarizing of the best practices of Miscanthus cultivation during multiyear production, including first year establishment phase, and harvest phase starting from the 2nd year till 20th year. The cultivation cost assessment with conversion to selected bioproducts is presented in Tables 9.1–9.3. The biggest investment is at the first year of plantation establishing from which the largest investment is for purchasing of rhizomes and labor cost. Another important cost driver is the field-to-farm distance. It was concluded (Winkler et al., 2020) that the implementation of Miscanthus into farming systems can be profitable in the following cases: (i) for lands with unfavorable conditions, such as awkward shapes, slopes, or low soil quality; (ii) for greening areas or soil protection corridor; (iii) when Miscanthus is directly used in own farm. Also profitable is to cultivate Miscanthus in smaller plantation up to 1 ha, and to utilize biomass for combustion, animal bedding, and anaerobic digestion, and from these three pathways production of animal bedding defined as the most reasonable (Winkler et al., 2020).

Miscanthus has a potential to play an important role to provide sustainably produced feedstock for the biogas sector (in short-term) and bioeconomy (in long-term) (Kiesel, 2020). It will assist to achieve net zero GHG emissions, since it can provide sustainably produced feedstock and a renewable carbon source for the chemical industry, which requires carbon for varieties of products. Miscanthus is suitable for biogas production, being more environmental in comparison with annual crops, may assist to decrease the environmental impact of the biogas sector. A market pull for Miscanthus biomass while utilized for biogas production may stimulate further investments into this crop. Development of integrated on-farm biorefineries for Miscanthus feedstock

TABLE 9.1

Miscanthus Production Cost When Biomass Is Utilized for Combustion, Animal Bedding, Biogas and Organic Biogas

Harvest Date	Utilization Pathway	Yield Level ($t\ DM\ ha^{-1}\ a^{-1}$)	Costs					Interest Rate	Total
			Machine	Material	Fertilizer	Energy	Labor		
First Year – Establishment ($€\ ha^{-1}\ a^{-1}$)									
March	Combustion	15 and 25	408	2931		120	245	38	3743
	Animal bedding								
October	Biogas		475	2408		142	249	43	3317
	Organic biogas								
Second Year – Harvest Phase ($€\ ha^{-1}\ a^{-1}$)									
March	Combustion	15	244	8	104	50	37	28	471
		25	291		173	69	51	32	624
	Animal bedding	15	414		104	51	88	43	708
		25	575		173	72	137	58	1023
October	Biogas	15	370		-	95	69	41	583
	Organic biogas								
	Biogas	25	498		-	141	107	54	808
	Organic biogas							41	795
Third Year – Harvest and Removal Phase ($€\ ha^{-1}\ a^{-1}$)									
March	Combustion	15	317	8	-	78	51	34	488
		25	361		-	96	63	38	566
	Animal bedding	15	487		-	78	102	50	725
		25	644		-	99	149	64	964
October	Biogas	15	400		-	110	75	43	636
	Organic biogas								
	Biogas	25	498		-	148	106	54	814
	Organic biogas								

Source: Modified from Winkler et al. (2020).
Miscanthus plantation = 1 ha.

TABLE 9.2

Miscanthus Production Cost When Biomass Is Utilized for Combustion, Animal Bedding, Biogas and Organic Biogas

Harvest Date	Utilization Pathway	Yield Level (t DM ha^{-1} a^{-1})	Costs						Total
			Machine	Material	Fertilizer	Energy	Labor	Interest Rate	
First Year – Establishment (€ ha^{-1} a^{-1})									
March	Combustion	15 and 25	356	2931		106	205	34	3632
	Animal bedding								
October	Biogas		413	2408		127	197	38	3183
	Organic biogas								
Second Year – Harvest Phase (€ ha^{-1} a^{-1})									
March	Combustion	15	199	8	104	45	35	22	413
		25	248		173	65	49	27	570
	Animal bedding	15	361		104	44	81	37	635
		25	522		173	65	130	51	949
October	Biogas	15	320		-	90	62	35	515
	Organic biogas								
	Biogas	25	452		-	140	94	48	742
	Organic biogas								

(Continued)

TABLE 9.2 (Continued)

Miscanthus Production Cost When Biomass Is Utilized for Combustion, Animal Bedding, Biogas and Organic Biogas

Harvest Date	Utilization Pathway	Yield Level (t DM ha^{-1} a^{-1})	Costs						Total
			Machine	Material	Fertilizer	Energy	Labor	Interest Rate	
Third Year – Harvest & Removal Phase (€ ha^{-1} a^{-1})									
March	Combustion	15	257	8	-	69	43	27	404
		25	303		-	89	55	31	486
	Animal bedding	15	419		-	69	89	42	627
		25	577		-	90	136	56	867
October	Biogas	335	400		-	102	61	35	541
	Organic biogas								
	Biogas	25	436		-	144	85	46	719
	Organic biogas								

Source: Modified from Winkler et al. (2020).
Miscanthus plantation = 10ha.

TABLE 9.3

Comparison of Profit While Utilizing Miscanthus Biomass

	Units	Combustion		Animal Bedding		Biogas		Organic Biogas	
		15[a]	25[a]	15[a]	25[a]	15[a]	25[a]	15[a]	25[a]
Plantation = 1 ha									
Annual Ø-production costs	€ ha⁻¹	635	777	860	1155	743	954	721	933
Biomass production costs	€ t⁻¹ a⁻¹	47	35	64	51	55	42	53	41
Methane costs	€ m⁻³					0.23	0.18	0.23	0.18
Min/max selling price	€ t⁻¹	65–95		106–600		1373 (field)		1373 (field)	
Min/max sales revenue	€ ha⁻¹ a⁻¹	975–1425	1625–2375	1590–9000	2,650–15,000	1373 (field)		1373 (field)	
Plantation = 10 ha									
Annual Ø-production costs	€ ha⁻¹	574	718	783	1155	671	884	648	862
Biomass production costs	€ t⁻¹ a⁻¹	42	32	58	51	50	39	48	38
Methane costs	€ m⁻³					0.21	0.17	0.20	0.16
Min/max selling price	€ t⁻¹	65–95		106–600		1373 (field)		1373 (field)	
Min/max sales revenue	€ ha⁻¹ a⁻¹	975–1425	1625–2375	1590–9000	2,650–15,000	1373 (field)		1373 (field)	

Source: Modified from Winkler et al. (2020).

Yield level (t DM ha⁻¹ a⁻¹).

utilization and biogas production may be an attractive business model for farmers which simultaneously reduces environmental impacts of agriculture by decreasing cultivation of annual crops (Kiesel, 2020).

9.8 Conclusions

The benefits of phytoremediation of contaminated sites include site improvement, carbon sequestration, and biomass products. Because of the growing importance of land, greater efforts should be made to improve soil quality and reclaim abandoned properties. Environmental, social, and economic values should be considered in applying a sustainable development approach to making decisions using phytoremediation with biomass production.

References

Bardos, R., Thomas, H., Smith, J., Harries, N., Evans, F., Boyle, R., Howard, T., Lewis, R., Thomas, A., & Haslam, A. (2018). The development and use of sustainability criteria in SuRF-UK's sustainable remediation framework. *Sustainability, 10*(6), 1781. https://doi.org/10.3390/su10061781.

Ben Fradj, N., Rozakis, S., Borzęcka, M., & Matyka, M. (2020). Miscanthus in the European bio-economy: A network analysis. *Industrial Crops and Products, 148,* 112281. https://doi.org/10.1016/j.indcrop.2020.112281.

Capuana, M. (2020). A review of the performance of woody and herbaceous ornamental plants for phytoremediation in urban areas. *IForest, 13*(2), 139–151. https://doi.org/10.3832/ifor3242-013.

Cattaneo, F., Barbanti, L., Gioacchini, P., Ciavatta, C., & Marzadori, C. (2014). 13C abundance shows effective soil carbon sequestration in Miscanthus and giant reed compared to arable crops under Mediterranean climate. *Biology and Fertility of Soils, 50*(7), 1121–1128. https://doi.org/10.1007/s00374-014-0931-x.

Communication from the Commission to the European Parliament, the Council, the European Economic and Social Committee and the Committee of the Regions. (2011). A roadmap for moving to a competitive low carbon economy in 2050. European Commission, Brussels, European Union.

Cosentino, S. L., Scordia, D., Testa, G., Monti, A., Alexopoulou, E., & Christou, M. (2018). The importance of perennial grasses as a feedstock for bioenergy and bioproducts. In Efthymia, A. (ed.), *Perennial Grasses for Bioenergy and Bioproducts,* 1–33. Elsevier, Amsterdam. https://doi.org/10.1016/b978-0-12-812900-5.00001-1.

Dondini, M., van Groenigen, K.-J., del Galdo, I., & Jones, M. B. (2009). Carbon sequestration under Miscanthus: A study of 13C distribution in soil aggregates. *GCB Bioenergy, 1*(5), 321–330. https://doi.org/10.1111/j.1757-1707.2009.01025.x.

Dubis, B., Jankowski, K. J., Załuski, D., Bórawski, P., & Szempliński, W. (2019). Biomass production and energy balance of *Miscanthus* over a period of 11 years: A case study in a large-scale farm in Poland. *GCB Bioenergy, 11*(10), 1187–1201. https://doi.org/10.1111/gcbb.12625.

Europe 2020. (2010). A strategy for smart, sustainable and inclusive growth. 35 p.

European Commission. (2019). *Towards a sustainable Europe by 2030.* 132 p. ISBN: 978-92-79-98963-6. https://doi.org/10.2775/676251.

Evangelou, M. W. H., Conesa, H. M., Robinson, B. H., & Schulin, R. (2012). Biomass production on trace element–contaminated land: A review. *Environmental Engineering Science, 29*(9), 823–839. https://doi.org/10.1089/ees.2011.0428.

Fiorenza, S., Oubre, C. L., & Ward, C. H. (1999). *Phytoremediation of Hydrocarbon-Contaminated Soils* (Vol. 2). CRC Press, Boca Raton, FL.

Geletukha, G. (2017). Development and prospects of bioenergy in Ukraine. *International Conference of Biomass for Energy 2017,* Kyiv, Ukraine, September 19–20.

Geletukha, G., Zheliezna, T., Tryboi, O., & Bashtovyi, A. (2016). Analysis of criteria for the sustainable development of bioenergy. *UABio Position Paper, 17,* 1–30.

Gillingham, K., & Stock, J. H. (2018). The cost of reducing greenhouse gas emissions. *Journal of Economic Perspectives, 32*(4), 53–72. https://doi.org/10.1257/jep.32.4.53.

Hansen, E. M., Christensen, B. T., Jensen, L. S., & Kristensen, K. (2004). Carbon sequestration in soil beneath long-term Miscanthus plantations as determined by 13C abundance. *Biomass and Bioenergy, 26*(2), 97–105. https://doi.org/10.1016/S0961-9534(03)00102-8.

Hastings, A. (2017). Economic and Environmental Assessment of Seed and Rhizome Propagated Miscanthus in the UK. *Frontiers in Plant Science, 8,* 1058. https://doi.org/10.3389/fpls.2017.01058.

Hein, L., Bagstad, K. J., Obst, C., Edens, B., Schenau, S., Castillo, G., Soulard, F., Brown, C., Driver, A., & Bordt, M. (2020). Progress in natural capital accounting for ecosystems. *Science, 367*(6477), 514–515. https://doi.org/10.1126/science.aaz8901.

Khanna, M., Dhungana, B., & Clifton-Brown, J. (2008). Costs of producing miscanthus and switchgrass for bioenergy in Illinois. *Biomass and Bioenergy, 32*(6), 482–493. https://doi.org/10.1016/j.biombioe.2007.11.003.

Kiesel, A. (2020). The potential of miscanthus as biogas feedstock. PhD diss., University of Hohenheim, Germany. 145 p.

Kvak, V., Stefanovska, T., Pidlisnyuk, V., Alasmary, Z., & Kharytonov, M. (2018). The long-term assessment of *Miscanthus × gigantheus* cultivation in the Forest-Steppe zone of Ukraine. *INMATEH-Agricultural Engineering, 54*(1), 113–120. https://doi.org/10.17707/AgricultForest.64.2.10.

Lal, R. (2016). Soil health and carbon management. *Food and Energy Security, 5*(4), 212–222. https://doi.org/10.1002/fes3.96.

Lewandowski, I., Clifton-Brown, J., Trindade, L. M., van der Linden, G. C., Schwarz, K.-U., Müller-Sämann, K., Anisimov, A., Chen, C.-L., Dolstra, O., Donnison, I. S., Farrar, K., Fonteyne, S., Harding, G., Hastings, A., Huxley, L. M., Iqbal, Y., Khokhlov, N., Kiesel, A., Lootens, P., & Kalinina, O. (2016). Progress on optimizing Miscanthus biomass production for the European bioeconomy: Results of the EU FP7 Project OPTIMISC. *Frontiers in Plant Science, 7*(NOVEMBER), 1620. https://doi.org/10.3389/fpls.2016.01620.

McCutcheon, S. C., & Schnoor, J. L. (2004). *Phytoremediation: Transformation and Control of Contaminants* (Vol. 121). John Wiley & Sons, Hoboken, NJ.

Mikhailova, E. A., Groshans, G. R., Post, C. J., Schlautman, M. A., & Post, G. C. (2019). Valuation of soil organic carbon stocks in the contiguous United States based on the avoided social cost of carbon emissions. *Resources, 8*(3), 153. https://doi.org/10.3390/resources8030153.

Ministry of Industry and Trade of the Czech Republic. (2018). Obnovitelné zdroje energie v roce 2017. Retrieved from: https://www.mpo.cz/assets/cz/energetika/statistika/obnovitelne-zdroje-energie/2018/12/Obnovitelne-zdroje-energie-v-roce-2017-new.pdf.

NASEM. (2019). *Negative Emissions Technologies and Reliable Sequestration: A Research Agenda.* National Academy of Sciences, Engineering and Medicine, The National Academies Press, Washington, DC.

Paustian, K., Larson, E., Kent, J., Marx, E., & Swan, A. (2019). Soil C sequestration as a biological negative emission strategy. *Frontiers in Climate, 1*(8), 8. https://doi.org/10.3389/fclim.2019.00008.

Pivetz, B. E. (2001). *Phytoremediation of contaminated soil and ground water at hazardous waste sites.* US Environmental Protection Agency EPA/540/S-01/500, Office of Research and Development.

REN21. (2020). *Renewables 2020 Global Status Report.* REN21 Secretariat, Paris. ISBN 978-3-948393-00-7.

RER 2030. Renewable Energy – Recast to 2030. (2019). *EU Science Hub: The European Commission's science and knowledge service.* European Union. https://ec.europa.eu/jrc/en/jec/renewable-energy-recast-2030-red-ii.

Roik, M. V., Sinchenko, V. M., Ivaschenko, O. O., Purkin, V. I., Kvak, V. M., Gumentik, M. Ya., Ganzhenko, O. M., Sabluk, V. T., Grischenko, O. M., Fuchilo, Ya. D., Goncharuk, G. S., Furman, V. A., Kocar, M. I., Cvigun, G. V., Kovalchuk, N. S., Nedyak, T. M., Vorozhko, S. P., Doronin, V. A., Driga, V. V., Buzinnii, M. V., Dubovii, U. P., Pedos, V. P., Balagura, O. V., Smirnih, V. M., Zaimenko, N. V., Rahmetov, D. B., Scherbakova, T. O., Rahmetov, S. D., & Katelevskii, V. M. (2019). *Miscanthus in Ukraine.* Kyiv, Ukraine: FOP Yamchinskiy Press. ISBN 978-617-7804-11-5 [in Ukrainian].

Rumpel, C., Amiraslani, F., Koutika, L.-S., Smith, P., Whitehead, D., & Wollenberg, E. (2018). Put more carbon in soils to meet Paris climate pledges. *Nature, 564*, 32–34.

Sherrington, C., Bartley, J., & Moran, D. (2008). Farm-level constraints on the domestic supply of perennial energy crops in the UK. *Energy Policy, 36*(7), 2504–2512. https://doi.org/10.1016/j.enpol.2008.03.004.

System of Environmental-Economic Accounting. (2014). *System of Environmental-Economic Accounting 2012: Experimental Ecosystem Accounting.* United Nations, New York.

Tryboi, O. V. (2018). Efficient biomass value chains for heat production from energy crops in Ukraine. *Energetika, 64*(2), 84–92. https://doi.org/10.6001/energetika.v64i2.3782.

Tsemekidi Tzeiranaki, S., Bertoldi, P., Paci, D., Castellazzi, L., Serrenho, T., Economidou, M., & Zangheri, P. (2020). *Energy consumption and energy efficiency trends in the EU-28, 2000–2018.* EUR 30328 EN, Publications Office of the European Union, Luxembourg, 156 p., ISBN 978-92-76-21074-0. https://doi.org/10.2760/847849. JRC120681.

U.S. EPA. 2011. *Handbook on the Benefits, Costs, and Impacts of Land Cleanup and Reuse.* U.S. Environmental Protection Agency, Washington, DC.

U.S. EPA. 2016. *The Social Cost of Carbon. EPA Fact Sheet, 2016.* U.S. Environmental Protection Agency, Washington, DC.

Von Cossel, M., Wagner, M., Lask, J., Magenau, E., Bauerle, A., Von Cossel, V., Warrach-Sagi, K., Elbersen, E., Staritsky, I., Van Eupen, M., Iqbal, Y., Jablonowski, N. D., Happe, S., Fernando, A. L., Scordia, D., Cosentino, S. L., Wulfmeyer, V., Lewandowski, I., & Winkler, B. (2019). Prospects of bioenergy cropping systems for a more social-ecologically sound bioeconomy. *Agronomy, 9*(10), 605. https://doi.org/10.3390/agronomy9100605.

Winkler, B., Mangold, A., von Cossel, M., Clifton-Brown, J., Pogrzeba, M., Lewandowski, I., Iqbal, Y., & Kiesel, A. (2020). Implementing miscanthus into farming systems: A review of agronomic practices, capital and labour demand. *Renewable and Sustainable Energy Reviews, 132*, 110053. https://doi.org/10.1016/j.rser.2020.110053.

Witzel, C. P., & Finger, R. (2016). Economic evaluation of Miscanthus production - A review. *Renewable and Sustainable Energy Reviews, 53*, 681–696. https://doi.org/10.1016/j.rser.2015.08.063.

Zappa, W., Junginger, M., & van den Broek, M. (2019). Is a 100% renewable European power system feasible by 2050? *Applied Energy, 233–234*, 1027–1050. https://doi.org/10.1016/j.apenergy.2018.08.109.

10

Miscanthus Biomass for Alternative Energy Production

Jikai Zhao, Donghai Wang, Valentina Pidlisnyuk, and Larry E. Erickson

Abstract

Biomass such as *Miscanthus* that is produced using phytoremediation can be used as a biofuel. In some locations it is used for heating homes that are located close to where it is produced. This chapter considers alternative energy technologies for *Miscanthus* and other plants, including liquid fuels such as ethanol, methane from anaerobic digestion, and pyrolysis (thermal Processing). Pretreatment alternatives to convert cellulose to glucose are reviewed because the economics of ethanol production from *Miscanthus* are impacted by the efficiency of cellulose hydrolysis to glucose. Size reduction of *Miscanthus* is often the first step in the process because enzymes for hydrolysis are more effective when there is large surface area. Ethanol that is produced by fermentation as a liquid is easier and less expensive to transport compared to *Miscanthus* biomass and methane from anaerobic digestion. Methane that is produced by anaerobic digestion is a mixture of methane and carbon dioxide. It can be used locally as a fuel, but it is expensive to distribute over a significant distance by pipeline. This chapter reviews pretreatment methods for *Miscanthus* prior to anaerobic digestion. There is a need to make good use of the hemicellulose and lignin because the economics are better when all of the harvested *Miscanthus* is used with good conversion efficiency.

CONTENTS

10.1 Introduction

The gradual depletion of nonrenewable fossil fuels and environmental deterioration due to the growing demand for energy sources and concern over greenhouse gas emissions have attracted considerable attention to exploring renewable and sustainable biofuels and supporting sustainable economic development (Arnoult & Brancourt-Hulmel, 2015). Lignocellulosic biomass, mainly composed of carbohydrate polymers (cellulose and hemicellulose) and an aromatic polymer (lignin), is widely identified as a promising alternative with great potential for biofuels production (Ho et al., 2019; Kim et al., 2016). Biological conversion of lignocellulosic biomass into ethanol, methane, hydrogen, heat, power, bio-oil, and syngas can reduce overdependence on petroleum-based fuels and mitigate climatic change (Brosse et al., 2012; Ge et al., 2016; Ziolkowska, 2014). In particular, bioethanol derived from lignocellulosic biomass has been utilized as a substitutive transportation biofuel to conventional gasoline (Bailey, 2018; von Blottnitz & Curran, 2007; Wyman, 2008).

The biomass of the second-generation crops including Miscanthus is processed to energy through distinct conversion routes: thermochemical and biochemical (Damartzis & Zabaniotou, 2011). The thermochemical route consists of the pyrolysis and/or gasification and subsequent gas cleaning and conditioning processes, followed by the Fischer–Tropsch synthesis for the production of synthetic liquid fuels. The biochemical route involves the enzymatic transformation of cellulose and hemicellulose to sugars and subsequent fermentation to bioethanol. The second route, although having more cost reduction potential due to its most recent development and constant effort for optimization, is less prone to commercialization than the first alternative. These two main pathways of biomass processing are illustrated in Figure 10.1.

The differences among the thermochemical processes are determined by the operation conditions of feed properties, oxidizer (air, oxygen or steam) amount, temperature, heating rate, and residence time.

10.2 Evaluation of Biomass Suitability for Energy

The key criteria for evaluating the suitability of plants as a raw material for combustion are the amount of biomass from 1 ha of cultivation, the amount of heat obtainable per unit weight of biomass, the cost of establishment of plantation, and the content of mineral substances determined as ash. The amount

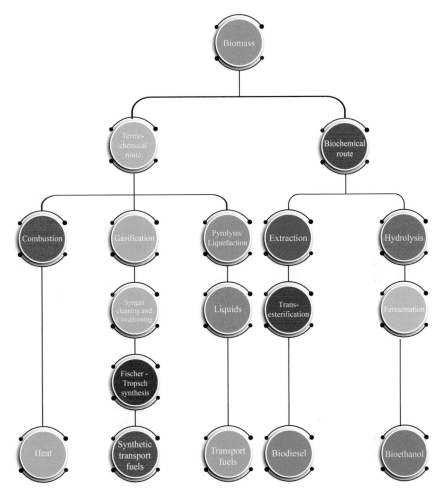

FIGURE 10.1
Schematic illustration of the main biofuel production pathways. (Modified from Damartzis & Zabaniotou, 2011.)

of biomass yield and the heating value for some grass plants are illustrated in Table 10.1.

Miscanthus as a rhizomatous C_4 perennial grass with low maintenance, rapid CO_2 absorption, significant carbon sequestration, and high biomass yield characteristics has been regarded as a dedicated energy crop for biofuels production, especially in Europe and North America (Ge et al., 2016; Hastings et al., 2009; Heaton et al., 2008; Lee & Kuan, 2015; Zub & Brancourt-Hulmel, 2010). The yield and chemical composition of Miscanthus biomass

TABLE 10.1

Biomass Yield from Grass Crops, Calorific Value, and the Cost of Cultivation

Plant Species	Biomass Yield (t ha^{-1})	Heating Value (MJ kg^{-1})
Tall wheatgrass	6.6–10.4	17.89
Tall oatgrass	7.5–12.4	18.29
Miscanthus	12.2–21.6	18.56

Source: Modified from Danielewicz et al. (2015).

are commonly influenced by the cultivation site, growing conditions, and harvest time (Arnoult et al., 2015; Kim et al., 2012; Le Ngoc Huyen et al., 2010), thus resulting in significant variation in bioconversion performance (Boakye-Boaten et al., 2016; Hodgson et al., 2010; Iqbal & Lewandowski, 2014). The heterogeneous nature of Miscanthus biomass allows its bioconversion into several added-value biofuels. For bioethanol production, pretreatment is an essential step to reduce the recalcitrance of biomass, rendering cellulose more amenable and accessible to enzymes (Lee & Kuan, 2015; Sun et al., 2016).

Anaerobic digestion and dark fermentation are usually used to convert Miscanthus biomass to biomethane and biohydrogen (de Vrije et al., 2009; Vasco-Correa & Li, 2015). During anaerobic digestion, anaerobic microbes can convert organic matter: pentose and hexose into biogas, methane and carbon dioxide (Frigon & Guiot, 2010). Dark fermentation is carried out under anaerobic conditions in which heterotrophic microorganisms degrade sugars by oxidation (Guo et al., 2010). The enzymatic attack of the microorganisms directly limits biomethane and biohydrogen production from lignocellulosic biomass. Appropriate pretreatment conditions are often required to accelerate conversion efficiency, including mechanical, thermochemical, and fungal methods (Frigon & Guiot, 2010; Guo et al., 2010). The potential of biogas production can also be affected by genotypes, harvesting time, and growing season (Mangold et al., 2019; Schmidt et al., 2018; Wahid et al., 2015).

At high temperatures, Miscanthus biomass can be subjected to thermochemical pretreatment to produce heat, power, bio-oil, and biogas that are compatible with current petrochemical infrastructures (Liu et al., 2017). Research findings indicate that operational temperature was the most influential factor in the yield and properties of bio-oil (Heo et al., 2010). Also, specific thermochemical reactors (fluidized bed, spouted bed, and fixed bed) assisted with catalytic and surfactant additives have been used to improve the conversion yield and quality of biofuels (Banks et al., 2014; Melligan et al., 2011; Yorgun & Şimşek, 2008). In this chapter, the biomass yield and chemical composition of Miscanthus biomass are summarized. The intrinsic mechanism of representative pretreatment methods used for bioethanol, biomethane, and biohydrogen production is thoroughly explained. Thermochemical conversion (combustion, pyrolysis, and gasification) of Miscanthus biomass to heat, power, bio-oil, and syngas is also presented. In addition, the internal and external factors that have significant influences on anaerobic digestion and thermochemical conversion performances of Miscanthus biomass are discussed. The flowchart for Miscanthus biomass conversion to different biofuels is illustrated in Figure 10.2.

The management practices for Miscanthus production (soil nutrient composition, amendments, irrigation, climate (precipitation, temperature)) are directly correlated with the properties of biomass and its potential for biofuel production (Cerazy-Waliszewska et al., 2019; Frydendal-Nielsen et al., 2016; Mangold et al., 2019; Wahid et al., 2015).

Representative studies on the chemical compositions of Miscanthus biomass are summarized in Table 10.2. Significant variations were identified in cellulose, hemicellulose, and lignin between the Miscanthus biomass samples, i.e., it is for cellulose 31.0–46.0%, for hemicellulose 13.6–35.4%, and for lignin 10.7–26.7%.

A comparison of chemical characteristics reveals differences in intrinsic genotypes, cultivation conditions, and harvesting times (Alam et al., 2019; Kim et al., 2012; Le Ngoc Huyen et al., 2010). Cellulose (D-glucose polymer) condenses through β (1–4) glycosidic bonds (Updegraff, 1969). Robust

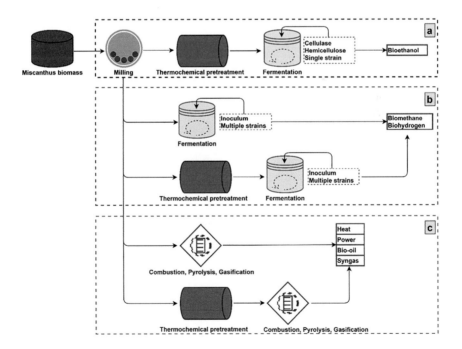

FIGURE 10.2
Flowchart of Miscanthus biomass conversion into biofuels: (a) bioethanol; (b) biomethane and biohydrogen; and (c) heat, power, bio-oil, and syngas.

TABLE 10.2

Chemical Composition of Miscanthus Biomass

Composition (%, Dry Basis)			
Cellulose	Hemicellulose	Lignin	Reference
46.0	27.8	10.7	Wang et al. (2010)
44.4	29.1	20.4	Alam et al. (2019)
44.3	30.3	21.7	Alam et al. (2019)
44.1	29.4	22.7	Alam et al. (2019)
43.3	13.6	26.3	Dash and Mohanty (2019)
43.1	23.6	26.3	Yang et al. (2015a)
41.2	21.2	25.1	Kang et al. (2013)
40.3	24.1	24.1	Cha et al. (2015b)
39.7	29.0	20.2	Alam et al. (2019)
39.5	30.5	22.0	Alam et al. (2019)
39.3	29.5	19.2	Alam et al. (2019)
39.2	23.5	21.4	Li et al. (2013)
38.6	17.9	25.4	Han et al. (2014)
38.0	18.5	20.9	Vasco-Correa et al. (2016)
37.2	30.9	21.9	Alam et al. (2019)
37.1	27.4	21.5	Alam et al. (2019)
37.0	22.1	23.3	Han et al. (2011)
36.3	22.8	21.3	Boakye-Boaten et al. (2015)
31.5	29.2	26.7	Si et al. (2015)
31.0	35.4	25.3	Si et al. (2015)
31.0	32.8	25.6	Si et al. (2015)

hydrogen bonds between and within cellulose strands are attributed to its high crystallinity. Miscanthus biomass is rich in cellulose (31.0%–46.0%) (Table 10.2). Taking into consideration that the removal of hemicellulose and lignin during the pretreatment process can lead to an approximate two-fold concentration of the remaining cellulose in pretreated biomass, high cellulose content in raw Miscanthus biomass would benefit the fermentable sugar concentration and final bioethanol titer. Hemicellulose (D-pentose polymer), a heterogeneous polysaccharide mix, is mainly composed of a β-D-xylose monomer in Miscanthus biomass, ranging from 13.6% to 35.4% (Table 10.2). Moreover, the hemicellulose is associated with the chemical and physical characteristics of subsequent biofuel. For example, the solubilization and elimination of hemicellulose are often critical to pretreatment effectiveness to increase enzymatic accessibility to cellulose (Zhao et al., 2020a). In the case of the lignin complex, it is randomly methoxylated and incorporated by lignols (*p*-coumaryl alcohol, coniferyl alcohol, and sinapyl alcohol). Lignin content in Miscanthus biomass is in the range of 10.7%–26.7%. Its lower free radicals make it more inert and could form nonproductive hydrophobic interaction with cellulase, thus reducing sugar and bioethanol yields.

10.3 Bioethanol Production

Several studies have reported using various pretreatment steps to reduce the recalcitrance of Miscanthus biomass for the valorization of macromolecules (Brosse et al., 2012; Ge et al., 2016). Miscanthus biomass is characterized by complex components, composing of cellulose, hemicellulose, and lignin (Zub & Brancourt-Hulmel, 2010). Therefore, an optimal combination of various pretreatment strategies is essential for efficient fractionation and further bioethanol production. The processing methods include mechanical treatment to improve the biomass maneuverability, thermochemical steps for the disruption and solubilization of unproductive compounds, and subsequent enzymatic hydrolysis and fermentation (Figure 10.1). The effectiveness of such processing procedures on the chemical composition, sugar recoveries, inhibitor formation, lignin removal, and bioethanol production performances is summarized and discussed in the following subsections.

10.3.1 Physicochemical Pretreatment

The reduction of particle size by mechanical chopping, grinding, or milling is often an initial pretreatment step of the solid starting feedstocks to facilitate subsequent thermochemical or enzymatic hydrolysis and fermentation disrupting their structural regularity and reducing the degree of crystallinity and polymerization (Hendriks & Zeeman, 2009). Generally, small particle sizes are preferred for efficient enzymatic hydrolysis due to their specific surface area, i.e., sugar accessibility. The particle size distribution is directly associated with chemical composition, sugar recoveries, delignification, and fermentation performances (Khullar et al., 2013). Generally, mechanical treatment by itself is incapable of disrupting and depolymerizing lignin that seals cellulose and hemicellulose tightly (Sun et al., 2016). Many thermochemical methods to conduct after preliminary size reduction have been investigated, including either the utilization of concentrated and dilute acid and alkali (Alam et al., 2019; Scordia et al., 2013; Si et al., 2015; Yoo et al., 2016; Zhao et al., 2020b), liquid hot water (LHW) and steam explosion (Li et al., 2013; Yeh et al., 2016), organosolv and ionic liquids (Brosse et al., 2009; Dash & Mohanty, 2019; Kim et al., 2018), or a combination of processing (Auxenfans et al., 2014; Rodríguez et al., 2011; Wang et al., 2010; Zhu et al., 2015). However, depending on the pretreatment conditions, various components might be formed as inhibitors that limit enzymatic activity.

Dilute acid (organic and inorganic acids) and alkali (metal hydroxide and aqueous ammonia) pretreatments have been extensively explored to enhance the enzymatic digestibility of Miscanthus biomass (Alam et al., 2019; Ji et al., 2015; Si et al., 2015; Vanderghem et al., 2012; Yoo et al., 2016). Since glucosidic bonds of cellulose and hemicellulose are susceptible to acid, a high proportion of hemicellulose and some of cellulose are hydrolyzed into slurries during pretreatment, and sugar degradation compounds such as furfural and hydroxymethylfurfural (HMF) as well as aromatic lignin degradation compounds can be generated (Mosier et al., 2005; Zhao et al., 2020a). Dilute acid pretreatment is commonly performed with high temperature (>150°C) and short time (<1 hour). Moreover, ethanol yield depends on pretreatment conditions applied (acid dose, reaction time, and temperature) (Ji et al., 2015). Alkali pretreatment can efficiently cleave and decompose the chemical cross-links (ether and ester bonds) between carbohydrates and lignin, resulting in a structural alteration of lignin and elimination of hemicellulose (Zhao et al., 2020b). The solubilization of disrupted lignin and hemicellulose renders Miscanthus biomass more amenable to enzymes due to the increment of cellulosic accessibility. Furthermore, alkali pretreatment is conducted under relatively low temperatures but long residence times, followed by multiple-washing for removing small lignin units and other inhibitors.

LHW and steam explosion as the category of hydrothermal pretreatments have attracted considerable attention for pretreatment of lignocellulosic biomass and solubilization of amorphous hemicellulose. During pretreatment, water can be autoionized into acidic hydronium ions that cleave the glycosidic bonds of hemicellulose, resulting in the formation of acetic acid, which in turn catalyzes breaking cellulose and hemicellulose into oligosaccharides and monomeric sugars (glucose and xylose) (Mosier et al., 2005). Thus, harsh pretreatment temperatures (180°C–230°C) induced further degradation and decomposition of monosaccharides into inhibitors (e.g., furfural and HMF) (Li et al., 2013). The formation of inhibitors in hydrolysates can cause sugar loss and inhibit subsequent enzymatic hydrolysis and fermentation. Thus, the detoxification process is commonly needed. Also, since the cellulose and lignin are more robust than hemicellulose, they are amenable for recovery. Therefore, hydrophobic interaction between residual lignin and cellulose during enzymatic saccharification is inevitable if there is no surfactant addition.

Organosolv and ionic liquids as green solvents offer the advantage of clean fractionation of lignocellulosic biomass into individual components with high purity (Brosse et al., 2009; Dash & Mohanty, 2019; Kim et al., 2018). Organosolv allows for the efficient fractionation of starting biomass into a solid residue rich in cellulose and a liquid fraction containing organosolv and water-soluble lignin and hemicellulose (Brosse et al., 2009). Ionic liquids owing hydrogen bond acceptor with high polarity can dissolve Miscanthus biomass, and ionic liquids having acetate, chloride, and phosphate anions show desirable solubility properties (Padmanabhan et al., 2011). However, excessive reagents are consumed for washing pretreated biomass to avoid lignin recondensation. Besides, the sealed condition required for organosolv and ionic liquid recoveries increases production costs, limiting their feasibility in commercialization.

To compensate for the drawbacks of a single pretreatment, the physical and chemical combinations such as microwave-assisted with acid and alkali

(Zhu et al., 2015), dilute acid presoaking coupled with wet explosion (Sørensen et al., 2008), dilute acid assisted with ionic liquids (Auxenfans et al., 2014), alkaline peroxide and electrolyzed water (Wang et al., 2010), ammonia with ionic liquids (Rodríguez et al., 2011), and aqueous ammonia with electron beam irradiation (Yang et al., 2015b) have been proposed to boost sugar and ethanol yields of Miscanthus biomass. Although some of these approaches, when combined, are potentially efficient for removing lignin and hemicellulose, complicated procedures need extra capital investment and operating costs. Besides, research only demonstrates their feasibility in the laboratory, more industrial or life cycle assessments coupled with detailed process economics are required before commercialization.

10.3.2 Enzymatic Hydrolysis and Fermentation

Apart from chemical compositions caused by external and intrinsic elements and pretreatment methods as discussed previously, solid and enzyme loading, surfactant addition, and applied microorganisms are also responsible for enzymatic hydrolysis and fermentation performances of biomass (Vanderghem et al., 2012).

Low solid loading would be beneficial for shortening enzymatic hydrolysis and fermentation duration and reaching high ethanol yield but causing low ethanol titer, which could be unable to meet the minimal requirement (around $40\,g\,L^{-1}$) for commercial ethanol distillation. Simultaneous saccharification and fermentation (SSF) of Miscanthus biomass was commonly conducted at solid loading less than 15% with the maximum ethanol titer less than $30\,g\,L^{-1}$ (Cha et al., 2015a; Scordia et al., 2013; Yoo et al., 2016). Given 60% of cellulose in pretreated Miscanthus biomass and 90% of glucose-to-ethanol conversion efficiency, the lowest solid loading for SSF should be greater than 13% to achieve $40\,g\,L^{-1}$ of ethanol concentration. Increasing solid loading within certain limits would theoretically enhance ethanol concentration. High solid loading with advantages of high ethanol titer and less water consumption is preferred from cost-efficiency and environmental standpoints (Chen et al., 2016). However, the decrease in glucan-to-ethanol yield is inevitable due to hydrophobic absorption between lignin and cellulase inhibiting enzymatic absorption and insufficient mixing (Kristensen et al., 2009). Several improvement strategies, such as surfactant addition (Alam et al., 2019), size reduction (Khullar et al., 2013), and thermotolerant microbial strains (Cha et al., 2015a), have been explored to enhance enzymatic saccharification and microbial digestion at high solid loading. However, initial studies aimed only to ferment glucose derived from cellulose using hexose-consuming microbial strains such as *Saccharomyces cerevisiae*. Recently, modified strains from genetic engineering capable of digesting pentose and hexose simultaneously and pentose-metabolizing bacterial strains of *Escherichia coli* have been developed to ferment the potential sugars in biomass sufficiently.

10.4 Biomethane and Biohydrogen Production

Biomethane and biohydrogen production is a renewable and sustainable technological process for Miscanthus biomass by anaerobic digestion (Figure 10.1). Compared to grain such as maize, Miscanthus-based biogas presents more economical and environmental advantages (Wagner et al., 2019).

Traditionally, Miscanthus biomass is first subjected to mechanical chopping and inoculated with anaerobic sludge performing SSF at thermophilic and mesophilic conditions (Klimiuk et al., 2010). The sewage or wastewater sludges, crop silages, and animal manures are commonly utilized as inoculum, which is composed of acetogenic bacteria and methanogens (Guo et al., 2010; Kiesel & Lewandowski, 2017; Klimiuk et al., 2010). During fermentation, biomacromolecules (cellulose and hemicellulose) are hydrolyzed to monomeric sugars (hexose and pentose) and then digested to organic acids and hydrogen by homoacetogens. The acetic acid and hydrogen generated as critical intermediates are rapidly consumed and transformed into CH_4 by methanogens (Guo et al., 2010). The potential of Miscanthus biomass for methane production has been reported to differ according to biomass harvest time, genotype, and plant fractions (Mangold et al., 2019; Schmidt et al., 2018; Wahid et al., 2015). This variation is associated with the compositional differences in the starting biomass, which can be mainly reflected by lignin incrustation that slowed down the enzymatic hydrolysis efficiency of polysaccharides (Klimiuk et al., 2010). In the case of fermentation, operational conditions, such as slurry pH, pressure, temperature, and microbial strains, also reflected notable differences in fermentation processes and were directly correlated with the conversion efficiency of carbohydrates to methane and hydrogen (Guo et al., 2010).

For biomethane production, physicochemical pretreatment methods have been often proposed to enhance biomass-to-biogas conversion efficiency by fractionation and decomposition of recalcitrant structures of Miscanthus biomass. Similar to bioethanol production, these mainly include size reduction, ensiling, steam explosion, LHW, acid, alkali, aqueous ammonia soaking, hydrogen peroxide, and enzymatic pretreatments (Jurado et al., 2013; Katukuri et al., 2017; Li et al., 2016; Menardo et al., 2013; Michalska et al., 2015; Nges et al., 2016; Zhou et al., 2017). However, the effectiveness of the pretreatment highly varied with pretreatment methods and conditions. For example, ensiled biomass showed higher methane yield and digestion rate than unensiled biomass (Mangold et al., 2019). Aqueous ammonia soaking increased methane yield by 25%–27% (Jurado et al., 2013), and hydrogen peroxide pretreatment increased methane yield by 49% (Katukuri et al., 2017). In the case of biohydrogen, one-step extrusion-NaOH pretreatment at moderate temperature resulted in 77% delignification and more than 95% of cellulose recovery as well as enhanced hydrogen yield (de Vrije et al., 2002). Mild alkali pretreatment assisted with *Caldicellulosiruptor saccharolyticus* and *Thermotoga neapolitana* strains to achieve hydrogen yields of 2.9 to 3.4 mol H_2 per mol of hexose, corresponding with 74%–85% of sugar yield (de Vrije et al., 2009). Although pretreatment functions to boost sugar conversion yield, effectiveness of the chosen pretreatment should be comprehensively determined by all-sided criteria, including sugar loss, inhibitor formation, sugar conversion yield, and capital input.

10.5 Thermochemical Conversion

Thermochemical conversion technologies, including combustion, pyrolysis, and gasification, are commonly utilized to produce heat, power, bio-oil, and syngas from lignocellulosic biomass (Liu et al., 2017; Saidur et al., 2011). The absence or presence of oxygen is the crucial difference between combustion and pyrolysis. The thermochemical conversion of Miscanthus biomass is discussed in the next section.

10.5.1 Heat and Power Generation

It has been reported that more than 90% of the world's bioenergy was obtained from the direct combustion of lignocellulosic biomass due to its high maneuverability and economic characteristics (Vassilev et al., 2013), as shown in Figure 10.1.

The calorific value is one of the most important parameters of biomass intended for use as a source of energy. The calorific value can be determined as higher heating value (HHV) which is the amount of heat released during fuel combustion when all products are turned back to precombustion state (25°C), so the heat of water condensation is included in value.

M. × giganteus elemental content and calorific value when the crop is produced at the regular agricultural land are summed up in Table 10.3 (Lewandowski et al., 2000) and Table 10.4 (Nebeska et al., 2019).

The use of Miscanthus biomass to produce heat and power, and the interconnection between combustion properties and agronomy practices (flowering, fertilization, senescence, and harvesting time), particle size, and genotypes are described (Baxter et al., 2012, 2014; Bilandzija et al., 2017; Clifton-Brown et al., 2004; Finnan & Burke, 2016; Iqbal & Lewandowski, 2016; Iqbal et al., 2017; Jensen et al., 2017; Lanzerstorfer, 2019; Meehan et al., 2013; Osman et al., 2017; Wilk et al., 2017). However, the presence of high quantities of alkali metal species in biomass ash often results in the formation of liquid phases such as alkali sulfates, silicates, and chlorides during combustion, which is responsible in slagging, fouling, corrosions, and agglomeration of bed material (Cruz et al., 2019; Morris et al., 2018; Nunes et al., 2016). In addition, the agglomeration severity is also related to operational variables, such

TABLE 10.3

M. × giganteus Biomass Elemental Content and Calorific Value

Country	Austria	Germany	Denmark	Greece
Age of stand (years)	3	3	3–5	2–3
Month of harvest	January/February	February/March	January/April	End of growing season
Water content (% fresh water)	30	16–28	21–48	38–44
Ash (% dry matter)	2.79	1.62–4.02	-	1.60
N (% dry matter)	0.49	0.19–0.39	0.58–0.67	0.33
P (% dry matter)	-	-		-
K (% dry matter)	-	0.52–0.94	0.31–0.48	-
S (% dry matter)	0.04	0.07–0.10	-	-
Cl (% dry matter)	0.24	0.10–0.17	0.04–0.12	-
C (% dry matter)	48.3	47.8–49.7	-	-
Calorific value (MJ kg^{-1})	19.12	17.05–18.54	-	-

Source: Modified from Lewandowski et al. (2000).

TABLE 10.4

The Calorific Value of Miscanthus Biomass While Produced in the Regular Agricultural Soil

Location of the Miscanthus Plantation	Harvesting Time	HHV (MJ kg^{-1})	Reference
Croatia	Autumn & winter & spring	18.19 ± 0.27[b]	Bilandzija et al. (2017)
France	n/d	17.80[b]	Jeguirim et al. (2010)
Germany	March	17.74	Michel et al. (2006)
Poland	July	19.04[b]	Dukiewicz et al. (2014)
Spain	n/d	18.07 ± 0.16[b]	García et al. (2012)
United Kingdom	October	17.5 4± 0.13[a]	Jensen et al. (2017)
United Kingdom	February	17.58 ± 0.05[a]	Jensen et al. (2017)
United Kingdom	September	18.20[a]	Mos et al. (2013)
United Kingdom	February	18.80[a]	Mos et al. (2013)
United Kingdom	February	19.19 ± 0.30[a]	Baxter et al. (2014)

Source: Modified from Nebeska et al. (2019).
[a] Calculation from ultimate analysis.
[b] Calorimetry.

as combustion temperature, fluidizing gas velocity, and additives (Morris et al., 2018). To date, ashes (bottom ash and fly ash) obtained after combustion are currently (around 70%) landfilled (Cruz et al., 2019). To valorize biomass ash, the biomass will be treated through hot water washing before the combustion process to remove the problematic chemical elements from the biomass efficiently and reduce acid gas formation and corrosion in the boiler (Gudka et al., 2016).

TABLE 10.5

Combustion Heats of *M. × giganteus* Biomass Produced at the Military Soil from Mimon (Czech Republic) and Control Soil

	HHV (MJ kg⁻¹)
Biomass from control soil	17.30 ± 0.2
Biomass from Mimon soil	17.10 ± 0.2
Other Fuels	
Dry wood (20%)	16.0
Brown coal (from Most, Czech Republic)	11.7–17.2
Black coal (from Ostrava, Czech Republic)	22.8–29.2
Coke	27.5
Mineral oil	40.6–42.3
Wheat straw	15.5
Paper	14,1
Waste plastics (separated)	23.0
Waste tires	25.0

Source: Modified from Nebeska et al. (2019).

In Table 10.5 the HHV value of *M. × giganteus* biomass produced at the military site (Mimon, Czech Republic) and control soil is presented (for sum: leaves+stems). The biomass was taken after 2 years of vegetation and harvested at the end of vegetation. In comparison, the amount of energy (combustion heat) obtained from burning of other fuels is summarized.

Even though the measuring was done for Miscanthus biomass that was not yet fully mature, the combustion energy was quite high, comparable to energy produced by wood or brown coal. When *M. × giganteus* was cultivated in Mimon soil, it had a slight negative effect on the value of combustion heat (17.10 ± 0.2 for biomass from Mimon military soil compared to 17.30 ± 0.2 for biomass from control soil).

10.5.2 Bio-Oil and Syngas Production

Pyrolysis and gasification have attracted considerable attention to convert Miscanthus biomass into liquid bio-oil, solid biochar, and syngas (carbon dioxide, carbon monoxide, hydrogen, and hydrocarbons) (Jayaraman & Gökalp, 2015). These are achieved by reacting the biomass at high temperatures (more than 400°C), without combustion, with a controlled amount of oxygen and steam, as shown in Figure 10.1. During pyrolysis and gasification, a number of chemical reactions are involved in the formation of bio-oil and syngas, including dehydration, depolymerization, isomerization, aromatization, decarboxylation, and charring (Kan et al., 2016; Wang et al., 2017), which occur chaotically as the observation of transitional behavior through thermogravimetric analysis (Jayaraman & Gökalp, 2015). The decomposition of biomass components was reported to vary: hemicellulose

(250°C–350°C), cellulose (325°C–400°C), and lignin (300°C–550°C) (Isahak et al., 2012; Kan et al., 2016). After fast pyrolysis, bio-oil, syngas, and bio-char typically account for 60%–70%, 13%–25%, and 12%–15%, respectively (Isahak et al., 2012).

The yield and quality of bio-oil and syngas during pyrolysis or gasification can be determined by parent biomass (organic composition, inorganic impurities, harvesting time, senescence time, and genotype) and operational (pretreatment, temperature, size reduction, feed rate, gas flow rate, and catalytic type) factors (Banks et al., 2014; Bok et al., 2013; Dickerson & Soria, 2013; Greenhalf et al., 2013; Heo et al., 2010; Kim et al., 2016; Kim et al., 2014; Melligan et al., 2011; Mos et al., 2013; Yorgun & Şimşek, 2008, 2003). Although biomass inorganics remain predominantly in the biochar, the fraction ejected as well as fine biochar particles entrained can have drastic impacts on the bio-oil properties and product yields (Liu et al., 2017). Inorganic impurities create specific challenges ranging from corrosion and fouling of surfaces to rapid and permanent deactivation of catalysts (Liu et al., 2017). Moreover, the chemical interaction between hemicellulose and lignin induces the generation of lignin-derived phenols, limiting hydrocarbon formation (Wang et al., 2011), and the cross-link between lignin and cellulose has a negative influence on the formation and distribution of pyrolysis products (Hosoya et al., 2007).

In order to make the biomass amenable, mechanical refining, thermal (torrefaction, steam explosion, ultrasound, and microwave irradiation), chemical (acid, alkali, and ionic liquid), and biological (microbial consortium and enzymes) pretreatments have been widely proposed to enhance biomass conversion efficiency and quality of bio-oil and syngas (Kan et al., 2016; Wang et al., 2017). However, these are not reviewed in this chapter due to the limited available studies. Regardless of which pretreatment is used, the intrinsic mechanism is mainly to (i) modify structural characteristics and alter chemical composition by decomposing hemicellulose and disrupting lignin; (ii) increase energy density; (iii) eliminate mineral substance (ash content). From the intrinsic mechanism, CO_2 and CO mainly derive from the degradation and recombination of carbonyl (C=O) and carboxyl (COO) groups (Qu et al., 2011), while CH_4 is primarily ascribed to methoxyl ($-O-CH_3$) and methylene ($-CH_2-$) groups and H_2 results from aromatic C=C and C–H groups (Liu et al., 2008). Therefore, changes in cellulose, hemicellulose, and lignin content of biomass would cause elemental variation, thus influencing the final composition and quality of bio-oil and syngas during pyrolysis.

The application of bio-oil obtained from biomass pyrolysis as candidate combustion fuels for electricity and heat production has been extensively investigated (Kan et al., 2016). However, due to the poor quality (weak volatility, high viscosity, and high corrosiveness), its long-term operation in a diesel test engine is still unfeasible (Bridgwater, 1999). Therefore, further bio-refining and upgrading of bio-oil using catalytic cracking technologies, high-pressure hydroprocessing, steam reforming, and gasification are needed to render it matchable to engines (Butler et al., 2011; Kan et al., 2016). For syngas, it can be directly used for gas combustion in spark ignition and compression ignition engines as well as further biofuel synthesis (Qu et al., 2011).

References

Alam, A., Zhang, R., Liu, P., Huang, J., Wang, Y., Hu, Z., Madadi, M., Sun, D., Hu, R., Ragauskas, A. J., Tu, Y., & Peng, L. (2019). A finalized determinant for complete lignocellulose enzymatic saccharification potential to maximize bioethanol production in bioenergy Miscanthus. *Biotechnology for Biofuels, 12*(1), 99. https://doi.org/10.1186/s13068-019-1437-4

Arnoult, S., & Brancourt-Hulmel, M. (2015). A review on Miscanthus biomass production and composition for bioenergy use: Genotypic and environmental variability and implications for breeding. *Bioenergy Research, 8*(2), 502–526. https://doi.org/10.1007/s12155-014-9524-7

Arnoult, S., Mansard, M.-C., & Brancourt-Hulmel, M. (2015). Early prediction of *Miscanthus* biomass production and composition based on the first six years of cultivation. *Crop Science, 55*(3), 1104–1116. https://doi.org/10.2135/cropsci2014.07.0493

Auxenfans, T., Buchoux, S., Husson, E., & Sarazin, C. (2014). Efficient enzymatic saccharification of Miscanthus: Energy-saving by combining dilute acid and ionic liquid pretreatments. *Biomass and Bioenergy, 62*, 82–92. https://doi.org/10.1016/j.biombioe.2014.01.011

Bailey, B. K. (2018). Performance of ethanol as a transportation fuel. In Charles E. Wyman (ed.), *Handbook on Bioethanol* (pp. 37–60). Routledge.

Banks, S. W., Nowakowski, D. J., & Bridgwater, A. V. (2014). Fast pyrolysis processing of surfactant washed Miscanthus. *Fuel Processing Technology, 128*, 94–103. https://doi.org/10.1016/j.fuproc.2014.07.005

Baxter, X. C., Darvell, L. I., Jones, J. M., Barraclough, T., Yates, N. E., & Shield, I. (2012). Study of *Miscanthus × giganteus* ash composition - Variation with agronomy and assessment method. *Fuel, 95*, 50–62. https://doi.org/10.1016/j.fuel.2011.12.025

Baxter, X. C., Darvell, L. I., Jones, J. M., Barraclough, T., Yates, N. E., & Shield, I. (2014). Miscanthus combustion properties and variations with Miscanthus agronomy. *Fuel, 117*(PART A), 851–869. https://doi.org/10.1016/j.fuel.2013.09.003

Bilandzija, N., Jurisic, V., Voca, N., Leto, J., Matin, A., Sito, S., & Kricka, T. (2017). Combustion properties of *Miscanthus × giganteus* biomass – Optimization of harvest time. *Journal of the Energy Institute, 90*(4), 528–533. https://doi.org/10.1016/j.joei.2016.05.009

Boakye-Boaten, N. A., Xiu, S., Shahbazi, A., & Fabish, J. (2015). Liquid hot water pretreatment of *Miscanthus × giganteus* for the sustainable production of bioethanol. *BioResources, 10*(3), 5890–5905.

Boakye-Boaten, N. A., Xiu, S., Shahbazi, A., Wang, L., Li, R., Mims, M., & Schimmel, K. (2016). Effects of fertilizer application and dry/wet processing of *Miscanthus × giganteus* on bioethanol production. *Bioresource Technology, 204*, 98–105. https://doi.org/10.1016/j.biortech.2015.12.070

Bok, J. P., Choi, H. S., Choi, J. W., & Choi, Y. S. (2013). Fast pyrolysis of *Miscanthus sinensis* in fluidized bed reactors: Characteristics of product yields and biocrude oil quality. *Energy, 60*, 44–52. https://doi.org/10.1016/j.energy.2013.08.024

Bridgwater, A. V. (1999). Principles and practice of biomass fast pyrolysis processes for liquids. *Journal of Analytical and Applied Pyrolysis, 51*(1), 3–22. https://doi.org/10.1016/S0165-2370(99)00005-4

Brosse, N., Dufour, A., Meng, X., Sun, Q., & Ragauskas, A. (2012). Miscanthus: A fast-growing crop for biofuels and chemicals production. *Biofuels, Bioproducts and Biorefining, 6*(5), 580–598. https://doi.org/10.1002/bbb.1353

Brosse, N., Sannigrahi, P., & Ragauskas, A. (2009). Pretreatment of *Miscanthus × giganteus* using the ethanol organosolv process for ethanol production. *Industrial and Engineering Chemistry Research, 48*(18), 8328–8334. https://doi.org/10.1021/ie9006672

Butler, E., Devlin, G., Meier, D., & McDonnell, K. (2011). A review of recent laboratory research and commercial developments in fast pyrolysis and upgrading. *Renewable and Sustainable Energy Reviews, 15*(8), 4171–4186. https://doi.org/10.1016/j.rser.2011.07.035

Cerazy-Waliszewska, J., Jeżowski, S., Łysakowski, P., Waliszewska, B., Zborowska, M., Sobańska, K., Ślusarkiewicz-Jarzina, A., Białas, W., & Pniewski, T. (2019). Potential of bioethanol production from biomass of various Miscanthus genotypes cultivated in three-year plantations in west-central Poland. *Industrial Crops and Products, 141*, 111790. https://doi.org/10.1016/j.indcrop.2019.111790

Cha, Y., An, G. H., Yang, J., Moon, Y., Yu, G., & Ahn, J. (2015a). Bioethanol production from Miscanthus using thermotolerant *Saccharomyces cerevisiae* MBC 2 isolated from the respiration-deficient mutants. *Renewable Energy, 80*, 259–265. https://doi.org/10.1016/j.renene.2015.02.016

Cha, Y., Yang, J., Park, Y., An, G. H., Ahn, J., Moon, Y., Yoon, Y., Yu, G., & Choi, I. (2015b). Continuous alkaline pretreatment of *Miscanthus sacchariflorus* using a bench-scale single screw reactor. *Bioresource Technology, 181*, 338–344. https://doi.org/10.1016/j.biortech.2015.01.079

Chen, X., Kuhn, E., Jennings, E. W., Nelson, R., Tao, L., Zhang, M., & Tucker, M. P. (2016). DMR (deacetylation and mechanical refining) processing of corn stover achieves high monomeric sugar concentrations ($230\,g\ L^{-1}$) during enzymatic hydrolysis and high ethanol concentrations (>10% v/v) during fermentation without hydrolysate purification or concentration. *Energy and Environmental Science, 9*(4), 1237–1245. https://doi.org/10.1039/c5ee03718b

Clifton-Brown, J. C., Stampfl, P. F., & Jones, M. B. (2004). Miscanthus biomass production for energy in Europe and its potential contribution to decreasing fossil fuel carbon emissions. *Global Change Biology, 10*(4), 509–518. https://doi.org/10.1111/j.1529-8817.2003.00749.x

Cruz, N. C., Silva, F. C., Tarelho, L. A. C., & Rodrigues, S. M. (2019). Critical review of key variables affecting potential recycling applications of ash produced at large-scale biomass combustion plants. *Resources, Conservation and Recycling, 150*, 104427. https://doi.org/10.1016/j.resconrec.2019.104427

Damartzis, T., & Zabaniotou, A. (2011). Thermochemical conversion of biomass to second generation biofuels through integrated process design-A review. *Renewable and Sustainable Energy Reviews, 15*(1), 366–378. https://doi.org/10.1016/j.rser.2010.08.003

Danielewicz, D., Surma-Slusarska, B., Żurek, G., & Martyniak, D., (2015). Selected grass plants as biomass fuels and raw material for papermaking. Part I. Caloritic valve and chemical composition. *BioResources, 10*(4), 8539–8551. https://doi.org/10.15375/biores.10.4.8539-8851

Dash, M., & Mohanty, K. (2019). Effect of different ionic liquids and anti-solvents on dissolution and regeneration of Miscanthus towards bioethanol. *Biomass and Bioenergy, 124*, 33–42. https://doi.org/10.1016/j.biombioe.2019.03.006

de Vrije, T., Bakker, R. R., Budde, M. A. W., Lai, M. H., Mars, A. E., & Claassen, P. A. M. (2009). Efficient hydrogen production from the lignocellulosic energy crop Miscanthus by the extreme thermophilic bacteria *Caldicellulosiruptor saccharolyticus* and *Thermotoga neapolitana*. *Biotechnology for Biofuels, 2*(1), 1–15. https://doi.org/10.1186/1754-6834-2-12

de Vrije, T., de Haas, G., Tan, G. B., Keijsers, E. R. P., & Claassen, P. A. M. (2002). Pretreatment of Miscanthus for hydrogen production by *Thermotoga elfii*. *International Journal of Hydrogen Energy, 27*(11–12), 1381–1390. https://doi.org/10.1016/S0360-3199(02)00124-6

Dickerson, T., & Soria, J. (2013). Catalytic fast pyrolysis: A review. *Energies, 6*(1), 514–538. https://doi.org/10.3390/en6010514

Dukiewicz, H., Waliszewska, B., & Zborowska, M. (2014). Higher and lower heating values of selected lignocellulose materials. *Annals of Warsaw University of Life Sciences – SGGW, Forestry and Wood Technology, 87*, 60–63.

Finnan, J., & Burke, B. (2016). Nitrogen fertilization of *Miscanthus × giganteus*: Effects on nitrogen uptake, growth, yield and emissions from biomass combustion. *Nutrient Cycling in Agroecosystems, 106*(2), 249–256. https://doi.org/10.1007/s10705-016-9793-y

Frigon, J. C., & Guiot, S. R. (2010). Biomethane production from starch and lignocellulosic crops: A comparative review. *Biofuels, Bioproducts and Biorefining, 4*(4), 447–458. https://doi.org/10.1002/bbb.229

Frydendal-Nielsen, S., Hjorth, M., Baby, S., Felby, C., Jørgensen, U., & Gislum, R. (2016). The effect of harvest time, dry matter content and mechanical pretreatments on anaerobic digestion and enzymatic hydrolysis of miscanthus. *Bioresource Technology, 218*, 1008–1015. https://doi.org/10.1016/j.biortech.2016.07.046

García, R., Pizarro, C., Lavín, A. G., & Bueno, J. L. (2012). Characterization of Spanish biomass wastes for energy use. *Bioresource Technology, 103*(1), 249–258. https://doi.org/10.1016/j.biortech.2011.10.004

Ge, X., Xu, F., Vasco-Correa, J., & Li, Y. (2016). Giant reed: A competitive energy crop in comparison with miscanthus. *Renewable and Sustainable Energy Reviews, 54*, 350–362. https://doi.org/10.1016/j.rser.2015.10.010

Greenhalf, C. E., Nowakowski, D. J., Yates, N., Shield, I., & Bridgwater, A. V. (2013). The influence of harvest and storage on the properties of and fast pyrolysis products from *Miscanthus × giganteus*. *Biomass and Bioenergy, 56*, 247–259. https://doi.org/10.1016/j.biombioe.2013.05.007

Gudka, B., Jones, J. M., Lea-Langton, A. R., Williams, A., & Saddawi, A. (2016). A review of the mitigation of deposition and emission problems during biomass combustion through washing pre-treatment. *Journal of the Energy Institute, 89*(2), 159–171. https://doi.org/10.1016/j.joei.2015.02.007

Guo, X. M., Trably, E., Latrille, E., Carrre, H., & Steyer, J. P. (2010). Hydrogen production from agricultural waste by dark fermentation: A review. *International Journal of Hydrogen Energy, 35*(19), 10660–10673. https://doi.org/10.1016/j.ijhydene.2010.03.008

Han, M., Kim, Y., Koo, B., & Choi, G. (2011). Bioethanol production by Miscanthus as a lignocellulosic biomass: Focus on high efficiency conversion to glucose and ethanol. *BioResources, 6*(2), 1939–1953.

Han, M., Moon, S. K., & Choi, G. W. (2014). Pretreatment solution recycling and high-concentration output for economical production of bioethanol. *Bioprocess and Biosystems Engineering, 37*(11), 2205–2213. https://doi.org/10.1007/s00449-014-1198-1

Hastings, A., Clifton-Brown, J., Wattenbach, M., Mitchell, C. P., Stampfl, P., & Smith, P. (2009). Future energy potential of Miscanthus in Europe. *GCB Bioenergy, 1*(2), 180–196. https://doi.org/10.1111/j.1757-1707.2009.01012.x

Heaton, E. A., Dohleman, F. G., & Long, S. P. (2008). Meeting US biofuel goals with less land: The potential of Miscanthus. *Global Change Biology, 14*(9), 2000–2014. https://doi.org/10.1111/j.1365-2486.2008.01662.x

Hendriks, A. T. W. M., & Zeeman, G. (2009). Pretreatments to enhance the digestibility of lignocellulosic biomass. *Bioresource Technology, 100*(1), 10–18. https://doi.org/10.1016/j.biortech.2008.05.027

Heo, H. S., Park, H. J., Yim, J. H., Sohn, J. M., Park, J., Kim, S. S., Ryu, C., Jeon, J. K., & Park, Y. K. (2010). Influence of operation variables on fast pyrolysis of *Miscanthus sinensis* var. purpurascens. *Bioresource Technology, 101*(10), 3672–3677. https://doi.org/10.1016/j.biortech.2009.12.078

Ho, M. C., Ong, V. Z., & Wu, T. Y. (2019). Potential use of alkaline hydrogen peroxide in lignocellulosic biomass pretreatment and valorization – A review. *Renewable and Sustainable Energy Reviews, 112*, 75–86. https://doi.org/10.1016/j.rser.2019.04.082

Hodgson, E. M., Fahmi, R., Yates, N., Barraclough, T., Shield, I., Allison, G., Bridgwater, A. V., & Donnison, I. S. (2010). Miscanthus as a feedstock for fast-pyrolysis: Does agronomic treatment affect quality? *Bioresource Technology, 101*(15), 6185–6191. https://doi.org/10.1016/j.biortech.2010.03.024

Hosoya, T., Kawamoto, H., & Saka, S. (2007). Cellulose-hemicellulose and cellulose-lignin interactions in wood pyrolysis at gasification temperature. *Journal of Analytical and Applied Pyrolysis, 80*(1), 118–125. https://doi.org/10.1016/j.jaap.2007.01.006

Iqbal, Y., Kiesel, A., Wagner, M., Nunn, C., Kalinina, O., Hastings, A. F. S. J., Clifton-Brown, J. C., & Lewandowski, I. (2017). Harvest time optimization for combustion quality of different Miscanthus genotypes across Europe. *Frontiers in Plant Science, 8*, 727. https://doi.org/10.3389/fpls.2017.00727

Iqbal, Y., & Lewandowski, I. (2014). Inter-annual variation in biomass combustion quality traits over five years in fifteen Miscanthus genotypes in south Germany. *Fuel Processing Technology, 121*, 47–55. https://doi.org/10.1016/j.fuproc.2014.01.003

Iqbal, Y., & Lewandowski, I. (2016). Biomass composition and ash melting behaviour of selected miscanthus genotypes in Southern Germany. *Fuel, 180*, 606–612. https://doi.org/10.1016/j.fuel.2016.04.073

Isahak, W. N. R. W., Hisham, M. W. M., Yarmo, M. A., & Yun Hin, T. Y. (2012). A review on bio-oil production from biomass by using pyrolysis method. *Renewable and Sustainable Energy Reviews, 16*(8), 5910–5923. https://doi.org/10.1016/j.rser.2012.05.039

Jayaraman, K., & Gökalp, I. (2015). Pyrolysis, combustion and gasification characteristics of miscanthus and sewage sludge. *Energy Conversion and Management, 89*, 83–91. https://doi.org/10.1016/j.enconman.2014.09.058

Jeguirim, M., Dorge, S., & Trouvé, G. (2010). Thermogravimetric analysis and emission characteristics of two energy crops in air atmosphere: *Arundo donax* and *Miscanthus giganteus*. *Bioresource Technology, 101*(2), 788–793. https://doi.org/10.1016/j.biortech.2009.05.063

Jensen, E., Robson, P., Farrar, K., Thomas Jones, S., Clifton-Brown, J., Payne, R., & Donnison, I. (2017). Towards *Miscanthus* combustion quality improvement: The role of flowering and senescence. *GCB Bioenergy, 9*(5), 891–908. https://doi.org/10.1111/gcbb.12391

Ji, Z., Zhang, X., Ling, Z., Zhou, X., Ramaswamy, S., & Xu, F. (2015). Visualization of *Miscanthus × giganteus* cell wall deconstruction subjected to dilute acid pretreatment for enhanced enzymatic digestibility. *Biotechnology for Biofuels, 8*(1), 1–14. https://doi.org/10.1186/s13068-015-0282-3

Jurado, E., Gavala, H. N., & Skiadas, I. V. (2013). Enhancement of methane yield from wheat straw, miscanthus and willow using aqueous ammonia soaking. *Environmental Technology (United Kingdom), 34*(13–14), 2069–2075. https://doi.org/10.1080/09593330.2013.826701

Kan, T., Strezov, V., & Evans, T. J. (2016). Lignocellulosic biomass pyrolysis: A review of product properties and effects of pyrolysis parameters. *Renewable and Sustainable Energy Reviews, 57*, 1126–1140. https://doi.org/10.1016/j.rser.2015.12.185

Kang, K. E., Han, M., Moon, S. K., Kang, H. W., Kim, Y., Cha, Y. L., & Choi, G. W. (2013). Optimization of alkali-extrusion pretreatment with twin-screw for bioethanol production from Miscanthus. *Fuel, 109*, 520–526. https://doi.org/10.1016/j.fuel.2013.03.026

Katukuri, N. R., Fu, S., He, S., Xu, X., Yuan, X., Yang, Z., & Guo, R. B. (2017). Enhanced methane production of *Miscanthus floridulus* by hydrogen peroxide pretreatment. *Fuel, 199*, 562–566. https://doi.org/10.1016/j.fuel.2017.03.014

Khullar, E., Dien, B. S., Rausch, K. D., Tumbleson, M. E., & Singh, V. (2013). Effect of particle size on enzymatic hydrolysis of pretreated Miscanthus. *Industrial Crops and Products, 44*, 11–17. https://doi.org/10.1016/j.indcrop.2012.10.015

Kiesel, A., & Lewandowski, I. (2017). Miscanthus as biogas substrate - cutting tolerance and potential for anaerobic digestion. *GCB Bioenergy, 9*(1), 153–167. https://doi.org/10.1111/gcbb.12330

Kim, J. S., Lee, Y. Y., & Kim, T. H. (2016). A review on alkaline pretreatment technology for bioconversion of lignocellulosic biomass. *Bioresource Technology, 199*, 42–48. https://doi.org/10.1016/j.biortech.2015.08.085

Kim, J. Y., Oh, S., Hwang, H., Moon, Y. H., & Choi, J. W. (2014). Assessment of miscanthus biomass (*Miscanthus sacchariflorus*) for conversion and utilization of bio-oil by fluidized bed type fast pyrolysis. *Energy, 76*, 284–291. https://doi.org/10.1016/j.energy.2014.08.010

Kim, S. J., Kim, M. Y., Jeong, S. J., Jang, M. S., & Chung, I. M. (2012). Analysis of the biomass content of various Miscanthus genotypes for biofuel production in Korea. *Industrial Crops and Products, 38*(1), 46–49. https://doi.org/10.1016/j.indcrop.2012.01.003

Kim, T., Im, D., Oh, K., & Kim, T. (2018). Effects of organosolv pretreatment using temperature-controlled bench-scale ball milling on enzymatic saccharification of *Miscanthus × giganteus*. *Energies, 11*(10), 2657. https://doi.org/10.3390/en11102657

Klimiuk, E., Pokój, T., Budzyński, W., & Dubis, B. (2010). Theoretical and observed biogas production from plant biomass of different fibre contents. *Bioresource Technology, 101*(24), 9527–9535. https://doi.org/10.1016/j.biortech.2010.06.130

Kristensen, J. B., Felby, C., & Jørgensen, H. (2009). Yield-determining factors in high-solids enzymatic hydrolysis of lignocellulose. *Biotechnology for Biofuels, 2*(1), 1–10. https://doi.org/10.1186/1754-6834-2-11

Lanzerstorfer, C. (2019). Combustion of Miscanthus: Composition of the ash by particle size. *Energies, 12*(1), 178. https://doi.org/10.3390/en12010178

Le Ngoc Huyen, T., Rémond, C., Dheilly, R. M., & Chabbert, B. (2010). Effect of harvesting date on the composition and saccharification of *Miscanthus* × *giganteus*. *Bioresource Technology, 101*(21), 8224–8231. https://doi.org/10.1016/j.biortech.2010.05.087

Lee, W.-C., & Kuan, W.-C. (2015). *Miscanthus* as cellulosic biomass for bioethanol production. *Biotechnology Journal, 10*(6), 840–854. https://doi.org/10.1002/biot.201400704

Lewandowski, I., Clifton-Brown, J. C., Scurlock, J. M. O., & Huisman, W. (2000). Miscanthus: European experience with a novel energy crop. *Biomass and Bioenergy, 19*(4), 209–227. https://doi.org/10.1016/S0961-9534(00)00032-5

Li, C., Liu, G., Nges, I. A., & Liu, J. (2016). Enhanced biomethane production from *Miscanthus lutarioriparius* using steam explosion pretreatment. *Fuel, 179*, 267–273. https://doi.org/10.1016/j.fuel.2016.03.087

Li, H. Q., Li, C. L., Sang, T., & Xu, J. (2013). Pretreatment on *Miscanthus lutarioriparious* by liquid hot water for efficient ethanol production. *Biotechnology for Biofuels, 6*(1), 1–10. https://doi.org/10.1186/1754-6834-6-76

Liu, Q., Chmely, S. C., & Abdoulmoumine, N. (2017). Biomass treatment strategies for thermochemical conversion. *Energy and Fuels, 31*(4), 3525–3536. https://doi.org/10.1021/acs.energyfuels.7b00258

Liu, Q., Wang, S., Zheng, Y., Luo, Z., & Cen, K. (2008). Mechanism study of wood lignin pyrolysis by using TG-FTIR analysis. *Journal of Analytical and Applied Pyrolysis, 82*(1), 170–177. https://doi.org/10.1016/j.jaap.2008.03.007

Mangold, A., Lewandowski, I., Hartung, J., & Kiesel, A. (2019). Miscanthus for biogas production: Influence of harvest date and ensiling on digestibility and methane hectare yield. *GCB Bioenergy, 11*(1), 50–62. https://doi.org/10.1111/gcbb.12584

Meehan, P. G., Finnan, J. M., & Mc Donnell, K. P. (2013). The effect of harvest date and harvest method on the combustion characteristics of *Miscanthus* × *giganteus*. *GCB Bioenergy, 5*(5), 487–496. https://doi.org/10.1111/gcbb.12003

Melligan, F., Auccaise, R., Novotny, E. H., Leahy, J. J., Hayes, M. H. B., & Kwapinski, W. (2011). Pressurised pyrolysis of Miscanthus using a fixed bed reactor. *Bioresource Technology, 102*(3), 3466–3470. https://doi.org/10.1016/j.biortech.2010.10.129

Menardo, S., Bauer, A., Theuretzbacher, F., Piringer, G., Nilsen, P. J., Balsari, P., Pavliska, O., & Amon, T. (2013). Biogas production from steam-exploded miscanthus and utilization of biogas energy and CO_2 in greenhouses. *Bioenergy Research, 6*(2), 620–630. https://doi.org/10.1007/s12155-012-9280-5

Michalska, K., Bizukojć, M., & Ledakowicz, S. (2015). Pretreatment of energy crops with sodium hydroxide and cellulolytic enzymes to increase biogas production. *Biomass and Bioenergy, 80*, 213–221. https://doi.org/10.1016/j.biombioe.2015.05.022

Michel, R., Mischler, N., Azambre, B., Finqueneisel, G., Machnikowski, J., Rutkowski, P., Zimny, T., & Weber, J. V. (2006). *Miscanthus* × *giganteus* straw and pellets as sustainable fuels and raw material for activated carbon. *Environmental Chemistry Letters, 4*(4), 185–189. https://doi.org/10.1007/s10311-006-0043-4

Morris, J. D., Daood, S. S., Chilton, S., & Nimmo, W. (2018). Mechanisms and mitigation of agglomeration during fluidized bed combustion of biomass: A review. *Fuel, 230*, 452–473. https://doi.org/10.1016/j.fuel.2018.04.098

Mos, M., Banks, S. W., Nowakowski, D. J., Robson, P. R. H., Bridgwater, A. V., & Donnison, I. S. (2013). Impact of *Miscanthus* × *giganteus* senescence times on fast pyrolysis bio-oil quality. *Bioresource Technology*, 129, 335–342. https://doi.org/10.1016/j.biortech.2012.11.069

Mosier, N., Wyman, C., Dale, B., Elander, R., Lee, Y. Y., Holtzapple, M., & Ladisch, M. (2005). Features of promising technologies for pretreatment of lignocellulosic biomass. *Bioresource Technology*, 96(6), 673–686. https://doi.org/10.1016/j.biortech.2004.06.025

Nebeska, D., Trogl, J., Zofkova, D., Voslarova, A., Stojdl, J., & Pidlisnyuk, V. (2019). Calorific values of *Miscanthus* × *giganteus* biomass cultivated under suboptimal conditions in marginal soils. *Studia Oecologica*, 13(1), 61–67. https://doi.org/10.21062/ujep/429.2020/a/1802-212X/SO/13/1/61

Nges, I. A., Li, C., Wang, B., Xiao, L., Yi, Z., & Liu, J. (2016). Physio-chemical pretreatments for improved methane potential of *Miscanthus lutarioriparius*. *Fuel*, 166, 29–35. https://doi.org/10.1016/j.fuel.2015.10.108

Nunes, L. J. R., Matias, J. C. O., & Catalão, J. P. S. (2016). Biomass combustion systems: A review on the physical and chemical properties of the ashes. *Renewable and Sustainable Energy Reviews*, 53, 235–242. https://doi.org/10.1016/j.rser.2015.08.053

Osman, A. I., Abdelkader, A., Johnston, C. R., Morgan, K., & Rooney, D. W. (2017). Thermal investigation and kinetic modeling of lignocellulosic biomass combustion for energy production and other applications. *Industrial and Engineering Chemistry Research*, 56(42), 12119–12130. https://doi.org/10.1021/acs.iecr.7b03478

Padmanabhan, S., Kim, M., Blanch, H. W., & Prausnitz, J. M. (2011). Solubility and rate of dissolution for Miscanthus in hydrophilic ionic liquids. *Fluid Phase Equilibria*, 309(1), 89–96. https://doi.org/10.1016/j.fluid.2011.06.034

Qu, T., Guo, W., Shen, L., Xiao, J., & Zhao, K. (2011). Experimental study of biomass pyrolysis based on three major components: Hemicellulose, cellulose, and lignin. *Industrial and Engineering Chemistry Research*, 50(18), 10424–10433. https://doi.org/10.1021/ie1025453

Rodríguez, H., Padmanabhan, S., Poon, G., & Prausnitz, J. M. (2011). Addition of ammonia and/or oxygen to an ionic liquid for delignification of miscanthus. *Bioresource Technology*, 102(17), 7946–7952. https://doi.org/10.1016/j.biortech.2011.05.039

Saidur, R., Abdelaziz, E. A., Demirbas, A., Hossain, M. S., & Mekhilef, S. (2011). A review on biomass as a fuel for boilers. *Renewable and Sustainable Energy Reviews*, 15(5), 2262–2289. https://doi.org/10.1016/j.rser.2011.02.015

Schmidt, A., Lemaigre, S., Ruf, T., Delfosse, P., & Emmerling, C. (2018). Miscanthus as biogas feedstock: Influence of harvest time and stand age on the biochemical methane potential (BMP) of two different growing seasons. *Biomass Conversion and Biorefinery*, 8(2), 245–254. https://doi.org/10.1007/s13399-017-0274-6

Scordia, D., Cosentino, S. L., & Jeffries, T. W. (2013). Effectiveness of dilute oxalic acid pretreatment of *Miscanthus* × *giganteus* biomass for ethanol production. *Biomass and Bioenergy*, 59, 540–548. https://doi.org/10.1016/j.biombioe.2013.09.011

Si, S., Chen, Y., Fan, C., Hu, H., Li, Y., Huang, J., Liao, H., Hao, B., Li, Q., Peng, L., & Tu, Y. (2015). Lignin extraction distinctively enhances biomass enzymatic saccharification in hemicelluloses-rich Miscanthus species under various alkali and acid pretreatments. *Bioresource Technology*, 183, 248–254. https://doi.org/10.1016/j.biortech.2015.02.031

Sørensen, A., Teller, P. J., Hilstrøm, T., & Ahring, B. K. (2008). Hydrolysis of Miscanthus for bioethanol production using dilute acid presoaking combined with wet explosion pre-treatment and enzymatic treatment. *Bioresource Technology, 99*(14), 6602–6607. https://doi.org/10.1016/j.biortech.2007.09.091

Sun, S., Sun, S., Cao, X., & Sun, R. (2016). The role of pretreatment in improving the enzymatic hydrolysis of lignocellulosic materials. *Bioresource Technology, 199,* 49–58. https://doi.org/10.1016/j.biortech.2015.08.061

Updegraff, D. M. (1969). Semimicro determination of cellulose inbiological materials. *Analytical Biochemistry, 32*(3), 420–424. https://doi.org/10.1016/S0003-2697(69)80009-6

Vanderghem, C., Brostaux, Y., Jacquet, N., Blecker, C., & Paquot, M. (2012). Optimization of formic/acetic acid delignification of *Miscanthus* × *giganteus* for enzymatic hydrolysis using response surface methodology. *Industrial Crops and Products, 35*(1), 280–286. https://doi.org/10.1016/j.indcrop.2011.07.014

Vasco-Correa, J., Ge, X., & Li, Y. (2016). Fungal pretreatment of non-sterile miscanthus for enhanced enzymatic hydrolysis. *Bioresource Technology, 203,* 118–123. https://doi.org/10.1016/j.biortech.2015.12.018

Vasco-Correa, J., & Li, Y. (2015). Solid-state anaerobic digestion of fungal pretreated *Miscanthus sinensis* harvested in two different seasons. *Bioresource Technology, 185,* 211–217. https://doi.org/10.1016/j.biortech.2015.02.099

Vassilev, S. V., Baxter, D., & Vassileva, C. G. (2013). An overview of the behaviour of biomass during combustion: Part I. Phase-mineral transformations of organic and inorganic matter. *Fuel, 112,* 391–449. https://doi.org/10.1016/j.fuel.2013.05.043

von Blottnitz, H., & Curran, M. A. (2007). A review of assessments conducted on bioethanol as a transportation fuel from a net energy, greenhouse gas, and environmental life cycle perspective. *Journal of Cleaner Production, 15*(7), 607–619. https://doi.org/10.1016/j.jclepro.2006.03.002

Wagner, M., Mangold, A., Lask, J., Petig, E., Kiesel, A., & Lewandowski, I. (2019). Economic and environmental performance of miscanthus cultivated on marginal land for biogas production. *GCB Bioenergy, 11*(1), 34–49. https://doi.org/10.1111/gcbb.12567

Wahid, R., Nielsen, S. F., Hernandez, V. M., Ward, A. J., Gislum, R., Jørgensen, U., & Møller, H. B. (2015). Methane production potential from *Miscanthus* sp.: Effect of harvesting time, genotypes and plant fractions. *Biosystems Engineering, 133,* 71–80. https://doi.org/10.1016/j.biosystemseng.2015.03.005

Wang, B., Wang, X., & Feng, H. (2010). Deconstructing recalcitrant Miscanthus with alkaline peroxide and electrolyzed water. *Bioresource Technology, 101*(2), 752–760. https://doi.org/10.1016/j.biortech.2009.08.063

Wang, S., Dai, G., Yang, H., & Luo, Z. (2017). Lignocellulosic biomass pyrolysis mechanism: A state-of-the-art review. *Progress in Energy and Combustion Science, 62,* 33–86. https://doi.org/10.1016/j.pecs.2017.05.004

Wang, S., Guo, X., Wang, K., & Luo, Z. (2011). Influence of the interaction of components on the pyrolysis behavior of biomass. *Journal of Analytical and Applied Pyrolysis, 91*(1), 183–189. https://doi.org/10.1016/j.jaap.2011.02.006

Wilk, M., Magdziarz, A., Gajek, M., Zajemska, M., Jayaraman, K., & Gokalp, I. (2017). Combustion and kinetic parameters estimation of torrefied pine, acacia and *Miscanthus* × *giganteus* using experimental and modelling techniques. *Bioresource Technology, 243,* 304–314. https://doi.org/10.1016/j.biortech.2017.06.116

Wyman, C. E. (2008). Cellulosic ethanol: A unique sustainable liquid transportation fuel. *MRS Bulletin, 33*(4), 381–383. https://doi.org/10.1557/mrs2008.77

Yang, F., Afzal, W., Cheng, K., Liu, N., Pauly, M., Bell, A. T., Liu, Z., & Prausnitz, J. M. (2015a). Nitric-acid hydrolysis of *Miscanthus* × *giganteus* to sugars fermented to bioethanol. *Biotechnology and Bioprocess Engineering, 20*(2), 304–314. https://doi.org/10.1007/s12257-014-0658-4

Yang, S. J., Yoo, H. Y., Choi, H. S., Lee, J. H., Park, C., & Kim, S. W. (2015b). Enhancement of enzymatic digestibility of Miscanthus by electron beam irradiation and chemical combined treatments for bioethanol production. *Chemical Engineering Journal, 275*, 227–234. https://doi.org/10.1016/j.cej.2015.04.056

Yeh, R. H., Lin, Y. S., Wang, T. H., Kuan, W. C., & Lee, W. C. (2016). Bioethanol production from pretreated *Miscanthus floridulus* biomass by simultaneous saccharification and fermentation. *Biomass and Bioenergy, 94*, 110–116. https://doi.org/10.1016/j.biombioe.2016.08.009

Yoo, H. Y., Yang, X., Kim, D. S., Lee, S. K., Lotrakul, P., Prasongsuk, S., Punnapayak, H., & Kim, S. W. (2016). Evaluation of the overall process on bioethanol production from miscanthus hydrolysates obtained by dilute acid pretreatment. *Biotechnology and Bioprocess Engineering, 21*(6), 733–742. https://doi.org/10.1007/s12257-016-0485-x

Yorgun, S., & Şimşek, Y. E. (2003). Fixed-bed pyrolysis of *Miscanthus* × *giganteus*: Product yields and bio-oil characterization. *Energy Sources, 25*(8), 779–790. https://doi.org/10.1080/00908310390207828

Yorgun, S., & Şimşek, Y. E. (2008). Catalytic pyrolysis of *Miscanthus* × *giganteus* over activated alumina. *Bioresource Technology, 99*(17), 8095–8100. https://doi.org/10.1016/j.biortech.2008.03.036

Zhao, J., Xu, Y., Wang, W., Griffin, J., & Wang, D. (2020a). Conversion of liquid hot water, acid and alkali pretreated industrial hemp biomasses to bioethanol. *Bioresource Technology, 309*, 123383. https://doi.org/10.1016/j.biortech.2020.123383

Zhao, J., Xu, Y., Zhang, M., & Wang, D. (2020b). Integrating bran starch hydrolysates with alkaline pretreated soft wheat bran to boost sugar concentration. *Bioresource Technology, 302*, 122826. https://doi.org/10.1016/j.biortech.2020.122826

Zhou, X., Li, Q., Zhang, Y., & Gu, Y. (2017). Effect of hydrothermal pretreatment on Miscanthus anaerobic digestion. *Bioresource Technology, 224*, 721–726. https://doi.org/10.1016/j.biortech.2016.10.085

Zhu, Z., Macquarrie, D. J., Simister, R., Gomez, L. D., & McQueen-Mason, S. J. (2015). Microwave assisted chemical pretreatment of Miscanthus under different temperature regimes. *Sustainable Chemical Processes, 3*(1), 1–13. https://doi.org/10.1186/s40508-015-0041-6

Ziolkowska, J. R. (2014). Prospective technologies, feedstocks and market innovations for ethanol and biodiesel production in the US. *Biotechnology Reports, 4*, 94–98. https://doi.org/10.1016/j.btre.2014.09.001

Zub, H. W., & Brancourt-Hulmel, M. (2010). Agronomic and physiological performances of different species of Miscanthus, a major energy crop. A review. *Agronomy for Sustainable Development, 30*(2), 201–214. https://doi.org/10.1051/agro/2009034

11

Miscanthus as Raw Materials for Bio-based Products

**Valentina Pidlisnyuk, Larry E. Erickson, Donghai Wang,
Jikai Zhao, Tatyana Stefanovska, and John R. Schlup**

Abstract

There is great interest in products from Miscanthus because large quantities of biomass can be produced annually. There are simple uses such as bedding for animals, mulch for horticulture applications, and insulation to improve energy conservation. Miscanthus has excellent natural absorbent qualities which makes it very attractive for spill management and as a bedding material. Compostable foodservice ware has been produced from Miscanthus to replace products from plastic that do not biodegrade. Building applications include fiberboard, particleboard, and composites. Miscanthus has high-quality cellulose for material applications and is an excellent source of cellulose where high quality is important. Nanocellulose applications from this crop are of interest and this is an active area of research, and cellulose from Miscanthus for paper production is one of the applications that is included in this chapter.

CONTENTS

11.1 Introduction

Miscanthus harvests provide large quantities of biomass. This chapter includes a review of alternative uses for Miscanthus that is produced at phytoremediation sites and regular agricultural lands. Since the locations and sizes of contaminated sites have diversity, it is good to have many potential uses for the biomass that is produced and harvested (Nsanganwimana et al., 2014). The interest in products that are biodegradable is growing because of pollution associated with plastic products that do not biodegrade.

The major constituents in Miscanthus are cellulose, hemicellulose, and lignin. These substances are very common in grasses, bushes, and trees. Wheat straw, corn stover, switchgrass, giant reed, reed canary grass, bamboo, and many other plants are good sources of cellulose, hemicellulose, and lignin. Because of all of the different sources, there is competition in the marketplace. There are also huge markets for biomass all over the world. The pulp and paper industry is an example. Biomass is used for energy, and it is better to burn renewable carbon than coal. The average breakdown of these components is shown in Table 11.1.

In the development of products from Miscanthus, the importance of making products from renewable biomass rather than from petrochemicals has received significant attention recently (Alexopoulou, 2018; Eschenhagen et al., 2019; Moll et al., 2020; Peças et al., 2018; Wang et al., 2018). Miscanthus removes carbon from the atmosphere as it grows; this carbon can be incorporated into building materials and other bio-based products, thus reducing the time associated with the carbon cycle. Carbon associated with sequestration is rapidly removed from the atmosphere and stored in bio-based products for many years.

Some products from Miscanthus are easy to implement and have use at many locations in numerous formulations. Examples of applications that require little processing of the Miscanthus include bedding for livestock and mulch in horticulture applications. It can also be employed as a soil

TABLE 11.1

Miscanthus Biomass Dry Weight Composition (wt. %)

Parameter	Value
Cellulose	42–50
Hemicellulose	25–28
Lignin	12–16
Ash	2.5–3
Water	10.5

amendment to add organic carbon and nutrients to a site. Like other types of biomass, Miscanthus pellets can be used as a fuel in stoves designed to burn pelletized wood and another biomass. This chapter will review the utilization of Miscanthus in bio-based materials.

11.2 Material Products

11.2.1 Agricultural Products

11.2.1.1 Bedding Applications

Bedding is an important resource for raising confined livestock. Due to its similarity to other agriculture materials employed in this application, over the last 10 years, there has been growing interest in Miscanthus as a high-quality product as bedding for poultry, pigs, sheep, cattle, and horses. One of the advantages of this application is that it has value in many locations. Its use in this application is straightforward, and it can be used directly without processing. It has good absorbent qualities, and it lasts longer than many other alternatives (Van Weyenberg et al., 2015). It is a clean product; the addition of the used bedding (Miscanthus with manure added) to soil has been shown to improve soil quality.

11.2.1.2 Mulch Applications

Mulch is used in many locations having diverse beneficial uses. Miscanthus may be used as a sustainable mulch in gardens and other horticulture applications such as in a new orchard (Samson et al., 2018). Mulch formed from the crop improves the aggregate stability of the soil to which it is applied. The increase in organic matter formation results in increases in the microbial numbers and earthworm populations. Phosphorus and potassium concentrations often increase while evaporation, weed growth, and soil erosion are reduced. This application is simple and it provides a good alternative for harvested Miscanthus in many countries.

11.2.2 Insulation

Miscanthus may be used for insulation in buildings to improve energy efficiency. A very clear example of this application was demonstrated in Wales with the construction of a house utilizing baled Miscanthus during 2017

(Construction Manager Magazine, 2017). Once framed, Miscanthus bales are placed within the frame in a manner similar to that employing wheat straw bales. The demonstration project was a collaboration between the University of Aberystwyth, the Centre for Alternative Technology, UK, and Terravesta (a company specializing in Miscanthus technology). In addition to taking advantage of Miscanthus as insulation, a primary goal is to reduce the large carbon footprint associated with conventional housing construction.

As an insulation material, harvested product may be used directly. However, there are several processing alternatives in using Miscanthus for insulation (Moll et al., 2020). A comprehensive review of insulation materials from biomass is available (Liu et al., 2017). Low thermal conductivity is important in insulation materials. In some applications sound absorption is very important as well. There is a growing interest in renewable materials because of the importance of reducing greenhouse gas emissions. This has resulted in an increase in the use of natural biomass insulation materials (Bozsaky, 2019) with a patent being issued for such applications (Huesemann-Lammert, 2006).

In 2012 Karl Schock, a German engineer, began the development of an insulating board based on Miscanthus or Napier grass. These are now fabricated through a German company, ISOCALM (GmbH, 2012), and are marketed as plaster carrier mats with thermal insulation. The reported thermal conductivity is 0.45 W m^{-1} K^{-1}.

Cardenas et al. (2015) demonstrated the use of Miscanthus as a block-type insulation material. While exploring formulations and processing conditions, thermal conductivities ranging between 0.079 and 0.116 W m^{-1} K^{-1} were obtained. The flexural strength of the Miscanthus product resembled that of expanded polystyrene, type IX.

Biomass insulation materials can be used in the natural form as some of the examples above illustrate. However, two obvious issues of concern with thermal insulation applications are fire safety and the impact of biological organisms using the Miscanthus as a food source. Any application as an insulation material must meet product standards related to fire safety. The integrity of the material must remain even if microbes and other organisms are present. Additives such as borax and aluminum sulfate may be added to improve fire resistance and reduce mold growth (Lopez Hurtado et al., 2016). Formulations have been reported which include ~1.75 wt.% borax, 1.75 wt.% sodium carbonate, and 0.5 wt.% fungicide.

The experience gained from formulating cellulose insulation with recycled paper may well be applicable. The thermal conductivity of cellulose insulation is approximately 0.040 W m^{-1} K^{-1}; that value increases with moisture content. Low moisture content is desirable as it does not support mold growth. Borate salts, boric acid, and aluminum sulfate are often mixed into cellulose fibers to prevent combustion and mold growth (Lopez Hurtado et al., 2016).

11.2.3 Composites, Building Materials, Cement

The construction industry relies on a diverse array of board products in which lignocellulose fibers and particles take a variety of forms. These products include plywood, particleboard, fiberboard, and strand board. All of these products typically involve the use of some kind of bonding agent. Typical fiberboard compositions are similar to 82% of wood/lignocellulose, 9% of resin, 1% of paraffin, and 8% of water.

Fiberboard and particleboard can be produced from Miscanthus and used as insulation and paneling (Tajuddin et al., 2016). Particleboard from this crop has good qualities when produced by hot pressing and steam processes. The data in Table 11.2 below provide data on a Miscanthus fiberboard fabricated by Velasquez et al. (2002). Property data for medium-density fiberboard is provided as a comparison.

Lignin has been used as an adhesive in the production of particleboard. It has a glass transition temperature of about 200°C and is an effective binder when particleboard is produced by hot pressing. Good results have been obtained using Miscanthus. This was the approach employed by Velasquez et al. (2002) and Salvadó et al. (2003). An example of hot-pressing conditions is 180°C and 5.3 MPa for 10 minutes, yielding a board thickness of 5 mm and product density of 1.0 g cm^{-3} (Tajuddin et al., 2016). In a comparison with other natural fibers, Miscanthus had the largest value of modulus of rupture at 61 MPa. In some work, additional lignin is added to improve bonding (Hubbe et al., 2018). While there has been significant progress in the development of insulation and building materials from Miscanthus biomass, particle size, the aspect ratios of the cellulose fibers, and bonding of particles are areas of ongoing research (Moll et al., 2020).

The development of lightweight concrete through the incorporation of biomass is not a new concept with a patent issued in 2008 with the assignee being Miscanthus-Holdings, SA (Luxembourg) (Hohn, 2008). Several studies of Miscanthus-based lightweight concrete have been reported (Chen et al., 2017, 2020; Ezechiels, 2017; Waldmann et al., 2016). In addition

TABLE 11.2

A Comparison of Properties for a Miscanthus Fiberboard versus Typical Data for Medium-Density Fiberboard

	Density (Mg m^{-3})	Specific Modulus (MPa kg^{-1}m^3)	Rupture Modulus (MPa)	Thickness Swelling	Water Adsorption	Reference
Miscanthus fiberboard	0.99–1.2	6.0	50–60	5%–60%	20%–40%	Velasquez et al. (2002)
Regular medium-density fiberboard	0.9–1.0	~6–8	60–90	-	-	CES Edupack (2017)

to producing a lower density concrete, formulations have been developed that have attractive acoustic absorption properties (Chen et al., 2017) and improved environmental impact (Courard & Parmentier, 2017). Typical densities are in the range of 0.65–1.25 Mg m^{-3}. A thermal conductivity of 0.17 W m^{-1} K^{-1} was reported at a density of 0.800 Mg m^{-3}. The goal has been to achieve compressive strengths exceeding 2.5 MPa; a 9 vol% fiber content has been shown to reduce the density by 20%. The volume content of the fiber in the mix is the important metric. It is important to realize that a fiber content of 9 vol% corresponds to 0.98 wt.% in the mix.

11.2.4 Composite Materials

The incorporation of Miscanthus fibers in composites has been investigated and the results have generated significant interest in developing products for use in buildings, vehicles, and other applications (Moll et al., 2020; Muthuraj et al., 2017). The amount of each component in the system, the dispersion of the fibers within the matrix, the particle size and aspect ratio of the Miscanthus fibers, the conditions of processing are important variables in producing composites for specific applications (Moll et al., 2020; Nagarajan et al., 2013). Materials in composites should be selected based on the desired properties for the application. The material properties of green composites have been reviewed and compared to those of other materials (Dicker et al., 2014). Biocomposites have the potential of lower costs and lower densities when compared to more traditional polymer matrix composites.

Miscanthus fibers have been incorporated into biocomposites. While any number of resin systems could be employed as the matrix, the materials selected for the matrix should be biodegradable if the biocomposite is to be truly biodegradable (Muthuraj et al., 2017). Polylactic acid, maleic anhydride, and poly(butylene succinate) (Muthuraj et al., 2015), poly(3-hydroxybutyrate-co-hydroxyvalerate) (Zhang et al., 2014), and biodegradable binary blends (Muthuraj et al., 2017) are examples of matrices used with Miscanthus fibers to form biocomposites.

Miscanthus fiber length has been found to be an important variable for impact strength (Muthuraj et al., 2016). Nanocellulose fibers have excellent properties when they are incorporated into composites (Barbash et al., 2019, 2020; Yang et al., 2019). The crop can be used as a raw material for the production of nanocellulose and cellulose nanocrystals (Cudjoe et al., 2017). Cellulose nanofibers may have crystalline domains, width between 3 and 20 nm, and length between 100 and 4000 nm (Yang et al., 2019). Nanocrystalline films have been formed with tensile strengths ranging from about 50 to 100 MPa (Barbash et al., 2020). The cellulose in Miscanthus has excellent quality for material applications, and it can be produced in large quantities at a reasonable price.

11.2.5 Hemicelluloses

Hydrothermal processes may be used to extract hemicelluloses from Miscanthus at temperatures from 160°C to 180°C (Wang et al., 2018; Xiao et al., 2017). Arabinoxylans are the most common hemicellulose polymers in Miscanthus (Schäfer et al., 2019). Hydrolysis of the hemicellulose yields mostly xylose and a small amount of arabinose (Schäfer et al., 2019). For some applications of cellulose, it is beneficial to remove most of the hemicellulose. There is a market for xylose as a raw material for fermentation and other uses.

11.3 Processing of Miscanthus to Fibers, Pulp, and Papers

Pulp for paper production is divided into three main categories (Liu et al., 2018): wood fiber, nonwood fiber, and waste paper; the shares of which in the pulp and paper industry are currently 63%, 3%, and 34%, respectively. Nevertheless, the increasing potential of nonwood resources is a consequence of decreasing of forest resources (Lwako et al., 2013). The overall potential of nonwood plant raw materials which can be processed to pulp is about 2.5 billion tons per year and is renewed annually (Saijonkari-Pahkala, 2001). The use of nonwood plant raw materials in the production of pulp goes back to the origins of the paper industry and remains an urgent focus for the industry as forestry resources diminish (Ververis et al., 2004). Promising alternative plants include different crops, and Miscanthus is a leading candidate (Danielewicz et al., 2018) due to its high growth rates, high lignin content, and ability to grow on marginal and degraded lands, where the food production is prohibited. This latter factor eliminates competition with food production. The pulp from perennial grass biomass can to be added to the existing wood feedstock (Bocianowski et al., 2019a, b), be processed separately (Danielewicz et al., 2015), or mixed with waste papers (Cappelletto et al., 2000).

Generally, the paper manufacturing process has several stages, i.e., raw material preparation and handling, pulp manufacturing, pulp washing and screening, chemical recovery, bleaching, stock preparation, and papermaking (Bajpai, 2018). Paper production is basically a two-step process in which a fibrous raw material is first converted into pulp, and then the pulp is converted into paper. The harvested wood is first processed so that the fibers are separated from the unusable fraction of the wood, the lignin. Pulp making can be done mechanically or chemically. The pulp is then bleached and further processed, depending on the type and grade of paper that is to be produced. In the paper factory, the pulp is dried and pressed to produce paper sheets. Post use, an increasing fraction of paper and paper products is recycled, which avoids landfilling or incinerating.

For papermaking, the internodes of Miscanthus stalks have better cellular composition than the nodes and pith. The stalks require less alkali for Kraft pulping compared to birch (Danielewicz et al., 2018; Kordsachia & Patt, 1991).

The technique used for the pulping operations consists of three procedures: preliminary dry-mechanical treatment which initially compresses the crop's stems, a high yield pulping process, followed by a peroxide bleach sequence. Cappelletto et al. (2000) argued that cells found in biomass, i.e., epithelial, sclerenchyma, parenchyma, and medullar rays do not possess the correct dimensions for papermaking. Hence, it is necessary to reduce the number of these cells prior to the pulping process by dry-mechanical treatment. This divides the raw material into uniform segments and to remove leaves and pith. In turn, this pretreatment decreases the chemical consumption during pulping, thus reducing pollution of wastewater. To generate papermaking pulps from fibrous fractions of Miscanthus, a four-stage mechanical pretreatment operation was designed (Cappelletto & Mongardini, 1997). The first stage compacted the Miscanthus stalks by cutting the raw material using a blade mill. The second step used fans to transport the cut material, and the third stage contained a cyclone that was utilized to divide light fractions. Finally, the last stage separated hefty fines. These operations help to obtain a higher number of fibers because foreign matter like dust, sand, and pebbles are removed; it eliminates useless material such as leaves, pith, epithelial, and parenchyma cells of stems. The resulting material will contain a higher percentage of fibers.

There are few standard methods for obtaining pulp from wood and non-wood raw materials, i.e., the sulfate, sulfite, and neutral-sulfite methods. These negatively influence the environment because of the application of sulfur-containing reagents for lignin removal from plant materials. Most of the lignin is removed during cooking, but there is some residual lignin that can be removed in an additional stage by bleaching, using chlorine-based and oxygen-based chemicals (Smook, 2002).

It has been shown that variations of the growing conditions may considerably influence the cooking results during pulping of *Miscanthus sinensis* (Kordsachia et al., 1993). Two different raw material samples were obtained from 3-year-old plantations in Germany and Sweden, and harvested in spring. The raw material grown in Germany gave better cooking results, in particular, higher pulping yields (Table 11.3).

TABLE 11.3

Comparison of Pulp Strength of Two Different Sources of *M. sinensis*

Country of Origin	Cooking Process	Breaking Length (km)	Tear Strength (cN)	Runability Factor
Germany	AS/AQ	8.32	70.8	7.6
	NS/AQ	8.12	61.8	7.1
Sweden	AS/AQ	7.80	74.0	7.6
	NS/AQ	7.71	80.8	7.9

Source: Modified from Kordsachia et al. (1993).

Cooking of the leaf-fraction results in much lower yield, brightness, and pulp viscosity in comparison with stalks (Kordsachia et al., 1993). However, when whole plant material is cooked, the adverse effect of leaves is hardly evident. Miscanthus sinensis has some outstanding features in comparison with other nonwood pulping raw materials, i.e., high delignification rate obtained with a low chemical change. The high yield, good bleaching ability, and excellent strength properties nearly match those of pulps prepared from poplar.

Organosolvent delignification has been suggested as an environmentally friendly process and an alternative way for obtaining pulp. Organic reagents have the potential to remove lignin and hemicelluloses at boiling temperatures. A variety of organic solvents such as esters, alcohols, ketones, and organic acids have been offered for cooking (Barbash et al., 2011). Among organic solvents, acetic acid is regarded as a potential agent to achieve extensive delignification due to its relatively low cost. The application of hydrogen peroxide during cooking promotes delignification of raw materials; increased brightness can also be achieved by delignification with peroxyl compounds. At the same time, less pronounced degradation of the cellulose is observed during cooking with such compounds. The cooking process is carried out at low temperature which helps to reduce energy consumption.

11.4 Production of Pulp from *M. × giganteus* Biomass Produced on Pb-Contaminated Soil

A laboratory experiment was done to evaluate the production of pulp from Miscanthus biomass growth in Pb contaminated soil with concentrations in between 583 and 604 mg kg^{-1}; other trace elements, Mn, Ni, Cu, Zn, Sr, Zr, were detected in smaller concentrations (Table 11.4). The biomass for production of pulp was harvested in spring 2018.

TABLE 11.4

Content of Trace Elements in the Soil of Three Replicated Plots, mg kg^{-1}

Trace Elements	Plot 1	Plot 2	Plot 3
Mn	452 ± 34.31	764.93 ± 50.32	468.19 ± 34.98
Fe	15,975 ± 92	16,949 ± 95	17,799 ± 97
Ni	16.29 ± 8.40	14.48 ± 8.65	17.34 ± 8.73
Cu	127.70 ± 5.83	134.80 ± 6.21	130.25 ± 6.20
Zn	130.15 ± 6.33	174.99 ± 7.18	146.42 ± 11.64
Sr	78.72 ± 2.08	80.72 ± 2.11	76.51 ± 2.16
Zr	665.84 ± 4.01	678.75 ± 4.08	660.72 ± 4.10
Pb	604.16 ± 9.60	612.01 ± 8.67	583.15 ± 8.83

Cooking of pulp from Miscanthus stalks was done by the peracetic method, which is environmentally friendly than traditional sulfate and sulfite methods of cellulose production and is characterized by lower energy costs compared with conventional and other organosolvent methods of delignification (Barbash et al., 2011, 2020).

The chemical composition of different parts of the Miscanthus stalks in comparison with other nonwood plants raw materials and hardwood and softwood species is given in Table 11.5.

It can be seen that according to the content of cellulose the mixture of Miscanthus stalks exceeds the content of cellulose in a mixture of wheat straw, rapeseed, hemp, wood; however, it is similar to a mixture of flax. Miscanthus stalks have a relatively high lignin content, close to the lignin content in rapeseed and spruce stems; have a close mineral content (ash

TABLE 11.5

Chemical Composition of Different Parts of Nonwood Plant Raw Materials and Wood, % from Mass of Absolutely Dry Raw Materials

| Parts Plants | Cellulose | Lignin | Solubility in | | RFW[a] | Ash |
			Water	NaOH		
M. × giganteus						
Mixture	53.3 ± 1.47	25.5 ± 0.645	3.3 ± 0.49	25.1 ± 0.79	1.86 ± 0.15	1.71 ± 0.14
Internodes	55.8 ± 1.48	25.1 ± 0.63	3.0 ± 0.48	24.1 ± 0.78	2.04 ± 0.16	1.60 ± 0.13
Knots	46.6 ± 1.39	27.0 ± 0.65	4.2 ± 0.51	27.9 ± 0.81	1.81 ± 0.17	1.77 ± 0.15
Wheat Straw						
Mixture	44.3 ± 1.33	16.5 ± 0.58	10.1 ± 0.5	38.4 ± 0.99	5.2 ± 0.2	6.6 ± 0.18
Stalk	46.2 ± 1.34	18.6 ± 0.60	6.0 ± 0.48	36.2 ± 0.98	4.6 ± 0.19	4.2 ± 0.17
Leaves	42.3 ± 1.35	15.2 ± 0.59	9.8 ± 0.52	40.1 ± 1.05	6.5 ± 0.19	9.4 ± 0.19
Rape						
Stalk	35.6 ± 1.28	22.9 ± 0.65	11.6 ± 0.52	25.6 ± 0.81	4.8 ± 0.19	3.3 ± 0.16
Root	28.3 ± 1.29	27.7 ± 0.71	10.9 ± 0.53	31.5 ± 0.82	2.4 ± 0.21	5.4 ± 0.18
Flax						
Mixture	59.6 ± 1.41	10.9 ± 0.58	4.1 ± 0.49	13.6 ± 0.77	4.7 ± 0.19	2.4 ± 0.14
Fiber	69.5 ± 1.52	6.1 ± 0.61	3.7 ± 0.43	13.4 ± 0.66	3.6 ± 0.11	1.5 ± 0.12
Wood part	42.0 ± 1.36	23.6 ± 0.73	5.2 ± 0.54	19.4 ± 0.82	5.2 ± 0.24	2.8 ± 0.15
Hemp						
Mixture	46.2 ± 1.33	17.0 ± 0.53	6.9 ± 0.49	25.0 ± 0.69	2.2 ± 0.13	2.6 ± 0.12
Bast	67.8 ± 1.51	6.5 ± 0.48	3.8 ± 0.47	20.8 ± 0.57	1.9 ± 0.12	1.5 ± 0.11
Wood part	42.2 ± 1.34	12.5 ± 0.69	5.1 ± 0.53	22.9 ± 0.72	3.7 ± 0.15	2.9 ± 0.14
Wood						
Birch tree	41.0 ± 1.29	21.0 ± 0.54	2.2 ± 0.52	11.2 ± 0.68	1.8 ± 0.15	0.5 ± 0.07
Spruce	46.1 ± 1.35	28.5 ± 0.61	7.3 ± 0.54	18.3 ± 0.59	2.9 ± 0.18	0.2 ± 0.05

Source: Modified from Barbash et al. (2018).

[a] RFW, resins, fats, waxes.

TABLE 11.6

Content of the Trace Elements in Pulp, mg kg^{-1}

Trace Elements	Sample 1	Sample 2	Sample 3	Average Value
Ti	3.64 ± 1.12	2.97 ± 1.08	3.53 ± 1.40	3.38 ± 1.20
Mn	2.24 ± 0.25	2.21 ± 0.23	2.96 ± 0.31	2.47 ± 0.26
Fe	16.02 ± 0.35	16.13 ± 0.33	25.02 ± 0.47	19.06 ± 0.38
Ni	0.24 ± 0.08	0.28 ± 0.08	0.29 ± 0.10	0.27 ± 0.09
Cu	88.77 ± 0.43	81.31 ± 0.39	104.67 ± 0.50	91.58 ± 0.44
Zn	23.14 ± 0.21	20.94 ± 0.19	28.55 ± 0.26	24.21 ± 0.22
Sr	0.22 ± 0.01	0.19 ± 0.01	0.24 ± 0.02	0.22 ± 0.01
Sn	0.71 ± 0.06	0.62 ± 0.06	0.76 ± 0.07	0.70 ± 0.06
Pb	13.06 ± 0.12	11.68 ± 0.12	15.08 ± 0.16	13.28 ± 0.13

content) to other nonwood plant materials; and significantly exceed the value of this indicator in softwood and hardwood (Danielewicz et al., 2015). The organosolvent peracetic pulp from *M. ×giganteus* with the duration of cooking 90 minutes was selected for investigation of physical properties. The strength properties of handmade sheets from peracetic pulps had the following physical and mechanical parameters: breaking length 8300 m, tear resistance 310 mN, burst resistance 220 kPa. The obtained data testify to high physical and mechanical indicators and the possibility of using this cellulose in the production of various types of paper and cardboard (Barbash et al., 2020).

This pulp was analyzed for the content of Pb and other elements; using the X-Ray Roentgen-fluorescence analysis, results are presented at Table 11.6. The concentrations of Mn, Fe, Ni, Cu, Zn Sr, Pb and Zr are limited, so pulp may be used for further processing.

The research describes the conversion process of *M. ×giganteus* biomass produced in trace elements contaminated soil to pulp using peracetic treatments. The delignification of initial raw material yielded pulp with a low lignin and ash content and high brightness at low energy costs and in a short cooking time with limited concentration of trace elements.

References

Alexopoulou, E. (2018). *Perennial Grasses for Bioenergy and Bioproducts: Production, Uses, Sustainability and Markets for Giant Reed, Miscanthus, Switchgrass, Reed Canary Grass and Bamboo.* Academic Press, London, UK.

Bajpai, P. (2018). Brief description of the pulp and papermaking process. In P. Bajpai (Ed.), *Biotechnology for Pulp and Paper Processing* (pp. 9–26). Springer, Singapore. https://doi.org/10.1007/978-981-10-7853-8_2.

Barbash, V. A., Poyda, V., & Deykun, I. (2011). Peracetic acid pulp from annual plants. *Cellulose Chemistry and Technology, 45*(9–10), 613–618. https://cellulosechemtechnol.ro/pdf/CCT45,9-10(2011)/p.613-618.pdf.

Barbash, V. A., Trembus, I., & Sokolovska, N. (2018). Performic pulp from wheat straw. *Cellulose Chemistry and Technology, 52*(7–8), 673–680. http://www.cellulosechemtechnol.ro/pdf/CCT7-8(2018)/p.673-680.pdf.

Barbash, V. A., Yashchenko, O. V., & Vasylieva, O. A. (2019). Preparation and properties of nanocellulose from *Miscanthus × giganteus*. *Journal of Nanomaterials, 2019*, 3241968. https://doi.org/10.1155/2019/3241968.

Barbash, V. A., Yashchenko, O. V., & Vasylieva, O. A. (2020). Preparation and application of nanocellulose from *Miscanthus × giganteus* to improve the quality of paper for bags. *SN Applied Sciences, 2*(4), 727. https://doi.org/10.1007/s42452-020-2529-2.

Bocianowski, J., Fabisiak, E., Joachimiak, K., & Wójciak, A. (2019a). NSSC pulping of miscanthus giganteus and birch wood Part 2: A comparison of papermaking potential and strength properties. *Wood Research, 64*(2), 281–291.

Bocianowski, J., Fabisiak, E., Joachimiak, K., Wojech, R., & Wojciak, A. (2019b). Miscanthus giganteus as an auxiliary raw material in NSSC birch pulp production. *Cellulose Chemistry and Technology, 53*(3–4), 271–279.

Bozsaky, D. (2019). Nature-based thermal insulation materials from renewable resources – A state-of-the-art review. *Slovak Journal of Civil Engineering, 27*(1), 52–59. https://doi.org/10.2478/sjce–2019-0008.

Cappelletto, P., & Mongardini, F. (1997). Industrial systems for preparation of cellulose fibers: IPZS experience. *Proceedings of the Flax and other plants Symposium*, Poznan, Poland.

Cappelletto, P., Mongardini, F., Barberi, B., Sannibale, M., Brizzi, M., & Pignatelli, V. (2000). Papermaking pulps from the fibrous fraction of *Miscanthus × giganteus*. *Industrial Crops and Products, 11*(2), 205–210. https://doi.org/10.1016/S0926-6690(99)00051-5.

Cardenas, J. P., Navia, R., Valdes, G., Zarrinbarkhsh, N., Misra, M., & Mohanty, A. K. (2015). Thermal insulation board based on Miscanthus residual fibers. *6th Annual Bioindustrial Meeting*, University of Alberta.

CES Edupack. (2017). Cambridge: Granta Design, Ltd.

Chen, Y., Wu, F., Yu, Q., & Brouwers, H. J. H. (2020). Bio-based ultra-lightweight concrete applying miscanthus fibers: Acoustic absorption and thermal insulation. *Cement and Concrete Composites, 114*, 103829. https://doi.org/10.1016/j.cemconcomp.2020.103829.

Chen, Y., Yu, Q. L., & Brouwers, H. J. H. (2017). Acoustic performance and microstructural analysis of bio-based lightweight concrete containing miscanthus. *Construction and Building Materials, 157*, 839–851. https://doi.org/10.1016/j.conbuildmat.2017.09.161.

Construction Manager Magazine. (2017). Welsh team build world's first house from miscanthus. *Construction Manager Magazine*. https://www.constructionmanagermagazine.com/welsh-team-build-house-miscanthus/.

Courard, L., & Parmentier, V. (2017). Carbonated miscanthus mineralized aggregates for reducing environmental impact of lightweight concrete blocks. *Sustainable Buildings, 2*(3), 9. https://doi.org/10.1051/sbuild/2017004.

Cudjoe, E., Hunsen, M., Xue, Z., Way, A. E., Barrios, E., Olson, R. A., Hore, M. J. A., & Rowan, S. J. (2017). *Miscanthus × giganteus*: A commercially viable sustainable source of cellulose nanocrystals. *Carbohydrate Polymers, 155*, 230–241. https://doi.org/10.1016/j.carbpol.2016.08.049.

Danielewicz, D., Dybka-Stępień, K., & Surma-Ślusarska, B. (2018). Processing of *Miscanthus × giganteus* stalks into various soda and kraft pulps. Part I: Chemical composition, types of cells and pulping effects. *Cellulose, 25*(11), 6731–6744. https://doi.org/10.1007/s10570-018-2023-9.

Danielewicz, D., Surma-Ślusarska, B., Żurek, G., Martyniak, D., Kmiotek, M., & Dybka, K. (2015). Selected grass plants as biomass fuels and raw materials for papermaking, Part II. Pulp and paper properties. *BioResources, 10*(4), 8552–8564. https://doi.org/10.15375/biores.10.4.8539-8851.

Dicker, M. P. M., Duckworth, P. F., Baker, A. B., Francois, G., Hazzard, M. K., & Weaver, P. M. (2014). Green composites: A review of material attributes and complementary applications. *Composites Part A: Applied Science and Manufacturing, 56*, 280–289. https://doi.org/10.1016/j.compositesa.2013.10.014.

Eschenhagen, A., Raj, M., Rodrigo, N., Zamora, A., Labonne, L., Evon, P., & Welemane, H. (2019). Investigation of Miscanthus and sunflower stalk fiber-reinforced composites for insulation applications. *Advances in Civil Engineering*, 1–7. https://doi.org/10.1155/2019/9328087.

Ezechiels, J. E. S. (2017). Design of an innovative bio-concrete using Miscanthus fibers [Master]. Eindhoven University of Technology.

GmbH. (2012). Telefonische Auskunft des Productionsleiters der Frima MEHA Dämmstoffe und Handels GmbH, Schifferstadt. www.isocalm.com.

Hohn, H. (2008). Method for producing concrete or mortar using a vegetal aggregate (United States Patent No. US7407615B2). https://patents.google.com/patent/US7407615B2/en.

Hubbe, M. A., Pizzi, A., Zhang, H., & Halis, R. (2018). Critical links governing performance of self-binding and natural binders for hot-pressed reconstituted lignocellulosic board without added formaldehyde: A review. *BioResources, 13*(1), 2049–2115. https://ojs.cnr.ncsu.edu/index.php/BioRes/article/view/BioRes_13_1_Hubbe_Review_Binders_Reconstituted_Lignocellulosic_Board.

Huesemann-Lammert, K. (2006). Natural fibers such as miscanthus, jute, flax or straw used as insulating material in the building industry (Germany Patent No. DE102004038050A1). https://patents.google.com/patent/DE102004038050A1/en.

Kordsachia, O., & Patt, R. (1991). Suitability of different hardwoods and non-wood plants for non-polluting pulp production. *Biomass and Bioenergy, 1*(4), 225–231. https://doi.org/10.1016/0961-9534(91)90007-Y.

Kordsachia, O., Seemann, A., & Patt, R. (1993). Fast growing poplar and *Miscanthus sinensis*—Future raw materials for pulping in Central Europe. *Biomass and Bioenergy, 5*(2), 137–143. https://doi.org/10.1016/0961-9534(93)90095-L.

Liu, L., Li, H., Lazzaretto, A., Manente, G., Tong, C., Liu, Q., & Li, N. (2017). The development history and prospects of biomass-based insulation materials for buildings. *Renewable and Sustainable Energy Reviews, 69*, 912–932. https://doi.org/10.1016/j.rser.2016.11.140.

Liu, Z., Wang, H., & Hui, L. (2018). Pulping and papermaking of non-wood fibers. In S. N. Kazi (Ed.), *Pulp and Paper Processing* (pp. 3–32). BoD – Books on Demand, Norderstedt, Germany. http://doi.org/10.5772/intechopen.79017.

Lopez Hurtado, P., Rouilly, A., Vandenbossche, V., & Raynaud, C. (2016). A review on the properties of cellulose fibre insulation. *Building and Environment, 96*, 170–177. https://doi.org/10.1016/j.buildenv.2015.09.031.

Lwako, M. K. O., Byaruhanga, J. K., & Baptist, K. J. (2013). A review on pulp manufacture from non-wood plant materials. *International Journal of Chemical Engineering and Applications, 4*(3), 144–148. https://www.cabdirect.org/cabdirect/abstract/20133301564.

Moll, L., Wever, C., Völkering, G., & Pude, R. (2020). Increase of Miscanthus cultivation with new roles in materials production—A review. *Agronomy, 10*(2), 308. https://doi.org/10.3390/agronomy10020308.

Muthuraj, R., Misra, M., Defersha, F., & Mohanty, A. K. (2016). Influence of processing parameters on the impact strength of biocomposites: A statistical approach. *Composites Part A: Applied Science and Manufacturing, 83*, 120–129. https://doi.org/10.1016/j.compositesa.2015.09.003.

Muthuraj, R., Misra, M., & Kumar Mohanty, A. (2017). Biocomposite consisting of miscanthus fiber and biodegradable binary blend matrix: Compatibilization and performance evaluation. *RSC Advances, 7*(44), 27538–27548. https://doi.org/10.1039/C6RA27987B.

Muthuraj, R., Misra, M., & Mohanty, A. K. (2015). Injection molded sustainable biocomposites from poly(butylene succinate) bioplastic and perennial grass. *ACS Sustainable Chemistry & Engineering, 3*(11), 2767–2776. https://doi.org/10.1021/acssuschemeng.5b00646.

Nagarajan, V., Mohanty, A. K., & Misra, M. (2013). Sustainable green composites: Value addition to agricultural residues and perennial grasses. *ACS Sustainable Chemistry & Engineering, 1*(3), 325–333. https://doi.org/10.1021/sc300084z.

Nsanganwimana, F., Pourrut, B., Mench, M., & Douay, F. (2014). Suitability of Miscanthus species for managing inorganic and organic contaminated land and restoring ecosystem services. A review. *Journal of Environmental Management, 143*, 123–134. https://doi.org/10.1016/j.jenvman.2014.04.027.

Peças, P., Carvalho, H., Salman, H., & Leite, M. (2018). Natural fibre composites and their applications: A review. *Journal of Composites Science, 2*(4), 66. https://doi.org/10.3390/jcs2040066.

Saijonkari-Pahkala, K. (2001). *Non-wood plants as raw material for pulp and paper* [Academic dissertation]. University of Helsinki. https://helda.helsinki.fi/bitstream/handle/10138/20756/nonwoodp.pdf?1.

Salvadó, J., Velásquez, J., & Ferrando, F. (2003). Binderless fiberboard from steam exploded *Miscanthus sinensis*: Optimization of pressing and pretreatment conditions. *Wood Science and Technology, 37*(3–4), 279–286. https://doi.org/10.1007/S00226-003-0186-4.

Samson, R., Delaquis, E., Deen, B., DeBruyn, J., & Eggimann, U. (2018). *A comprehensive guide to switchgrass management.* Ontario Biomass Producers Co-Operative Inc. http://www.ontariobiomass.com/resources/Documents/KTT%20Projects/KTT%20Documents%20and%20Videos/SwitchgrassFinal.pdf.

Schäfer, J., Sattler, M., Iqbal, Y., Lewandowski, I., & Bunzel, M. (2019). Characterization of Miscanthus cell wall polymers. *GCB Bioenergy, 11*(1), 191–205. https://doi.org/10.1111/gcbb.12538.

Smook, G. A. (2002). In M. J. Kocurek (ed.), *Handbook for Pulp and Paper Technologists* (425, 3rd ed.), Angus Wilde Publications Inc., Vancouver. ISBN 0-9694628-5-9.

Tajuddin, M., Ahmad, Z., & Ismail, H. (2016). A review of natural fibers and processing operations for the production of binderless boards. *BioResources, 11*(2), 5600–5617. https://ojs.cnr.ncsu.edu/index.php/BioRes/article/view/BioRes_11_2_Review_Tajuddin_Natural_Fibers_Processing_Operations_Binderless.

Van Weyenberg, S., Ulens, T., De Reu, K., Zwertvaegher, I., Demeyer, P., & Pluym, L. (2015). Feasibility of Miscanthus as alternative bedding for dairy cows. *Veterinarni Medicina, 60*(3). https://doi.org/10.17221/8058-vetmed.

Velasquez, J., Ferrando, F., & Salvadó, J. (2002). Binderless fiberboard from steam exploded *Miscanthus sinensis:* The effect of a grinding process. *Holz Als Roh-Und Werkstoff, 60*(4), 297–302. https://doi.org/10.1007/S00107-002-0304-2.

Ververis, C., Georghiou, K., Christodoulakis, N., Santas, P., & Santas, R. (2004). Fiber dimensions, lignin and cellulose content of various plant materials and their suitability for paper production. *Industrial Crops and Products, 19*(3), 245–254. https://doi.org/10.1016/j.indcrop.2003.10.006.

Waldmann, D., Thapa, V., Dahm, F., & Faltz, C. (2016). Masonry blocks from lightweight concrete on the basis of Miscanthus as aggregates. In S. Barth, D. Murphy-Bokern, O. Kalinina, G. Taylor, & M. Jones (Eds.), *Perennial Biomass Crops for a Resource-Constrained World* (pp. 273–295). Springer International Publishing, Cham, Switzerland.

Wang, K.-T., Jing, C., Wood, C., Nagardeolekar, A., Kohan, N., Dongre, P., Amidon, T. E., & Bujanovic, B. M. (2018). Toward complete utilization of Miscanthus in a hotwater extraction-based biorefinery. *Energies, 11*(1), 39. https://doi.org/10.3390/en11010039.

Xiao, L.-P., Song, G.-Y., & Sun, R.-C. (2017). Effect of hydrothermal processing on hemicellulose structure. In H. A. Ruiz, M. Hedegaard Thomsen, & H. L. Trajano (Eds.), *Hydrothermal Processing in Biorefineries: Production of Bioethanol and High Added-Value Compounds of Second and Third Generation Biomass* (pp. 45–94). Springer International Publishing, New York. https://doi.org/10.1007/978-3-319-56457-9_3.

Yang, H., Zhang, Y., Kato, R., & Rowan, S. J. (2019). Preparation of cellulose nanofibers from *Miscanthus × giganteus* by ammonium persulfate oxidation. *Carbohydrate Polymers, 212*, 30–39. https://doi.org/10.1016/j.carbpol.2019.02.008.

Zhang, K., Nagarajan, V., Zarrinbakhsh, N., Mohanty, A. K., & Misra, M. (2014). Co-injection molded new green composites from biodegradable polyesters and miscanthus fibers. *Macromolecular Materials and Engineering, 299*(4), 436–446. https://doi.org/10.1002/mame.201300189.

12

Conclusions and Recommendations

Larry E. Erickson, Valentina Pidlisnyuk, and Lawrence C. Davis

Abstract

Contamination of soil is a global concern. All inhabited parts of the world have land that is not being used productively because of past activities that have reduced soil quality. Improving the effectiveness and efficiency of phytotechnologies as well as reducing the costs of site restoration have value for society as better methods are used for revitalization of the contaminated sites. Great progress in developing phytotechnologies with biomass production has been reported. Miscanthus has been used with good results at many sites because the plant grows well in different marginal and contaminated soils. New knowledge has been developed on actions to take to increase biomass production at contaminated sites. The benefits of harvesting useful products and improving soil health during Miscanthus growth including those with adding amendments positively influenced the economics of land reclamation using phytotechnologies. Increased efforts to apply phytotechnologies with biomass production to improve contaminated sites and processing of harvested Miscanthus to different bioproducts are recommended.

CONTENTS

12.1 Conclusions

The applications of phytotechnologies depend on climate, local conditions including soil health, and plans for the site. The plans to address a contamination issue at a specific site should be developed based on available information on the site and on science and engineering knowledge. Miscanthus is one of many plants that have value for phytoremediation applications and has been investigated in several studies with positive results.

The contents of this book and in the references that have been included provide very useful information. Phytoremediation with biomass production has value as sustainability of land use is considered at specific sites that have contamination from past military and mining activities or other uses. Social value, environmental goals, and economic benefits enter into consideration as plans are developed. There has been very significant scientific progress during the last 30 years and many successful field applications have been carried out as reported in this book.

Chapters 2 and 3 review progress in developing phytotechnologies for addressing inorganic and organic soil contaminants, respectively. Miscanthus has been grown in many different environments with good results. Energy crops have been used in soils with trace elements contamination as well as at sites with organic contaminants. Phytostabilization with biomass production has been developed and applied at sites with inorganic contaminants such as lead and other trace elements. Methods to reduce the bioavailability of the trace elements are beneficial where biomass production is the primary objective and the uptake of trace elements into the biomass is intended to be small. There are many biomass products such as wood and paper where small quantities of trace elements in the finished product are acceptable with respect to health and safety. National and international regulations of course must be observed in this regard.

For soils with organic contaminants, the better options use strategies in which the contaminants are biodegraded or transformed by chemical processes. The goal is often to improve soil health such that the concentrations of any contaminants are reduced to levels so that they are no longer a concern, or are converted into less harmful products. Phytoremediation with biomass production can be implemented at many sites with organic contaminants where the biomass products are cellulosic fiber rather than food or feed products.

The economics associated with establishing and growing vegetation at contaminated sites can be improved if a trace element that has commercial value can be extracted from the soil by plants. Chapter 4 addresses to phytomining – the process of extracting a product such as nickel from soil using hyperaccumulator plants that are able to grow in the contaminated soil and accumulate a element product of value. Typically, after harvesting the plant biomass, drying it, and burning it for energy recovery, the ash can be processed to extract the element of interest.

The establishment of vegetation at some field sites requires a significant effort because of the physical state of the site, soil quality, and toxicity of the contaminants. Chapter 5 provides information on the science and technology for best practices when establishing vegetation at a contaminated site. Soil amendments, tilling, plant selection and production, water and weed management, harvest timing and nutrient optimization considerations are reviewed. Microbial ecology is altered by soil amendments and tilling, often in a positive way. Chapter 7 reviews biological options at contaminated sites and the benefits of adding microbial populations. Plant–microbe associations act to improve the growth of vegetation, and overall soil health.

In applications of phytoremediation with biomass production, it is useful to have multiple approaches to reduce the effects of the contamination on soil health and to improve soil health and biomass productivity. Chapter 6 reviews processes to improve soil health and enhance ecosystem services. Soil amendments that add organic carbon and living organisms may help to improve soil health, plant growth, and nutrient cycling.

Plant-feeding insects and nematodes have the potential to impact Miscanthus growth and product yield. It is important to consider pest migration from one crop to another when fields are nearby. Chapter 8 provides information on important plant-feeding insects and nematodes that have been found in Miscanthus plantations and have been studied and reported in published literature, including Miscanthus mealybugs, aphids, May beetles, plant parasitic nematodes, armyworms, and rootworms.

A forward-looking approach in economic aspects of phytoremediation with biomass production is addressed in Chapter 9. This is followed by alternative ways to convert Miscanthus biomass to energy (Chapter 10) and to different bioproducts (Chapter 11). Economic aspects are introduced with full consideration of environmental, social, ecosystem, and economic benefits associated with the improvement of each contaminated site. A sustainable remediation approach is used along with options for increasing the value chain of Miscanthus. The benefits of soil remediation with biomass production include risk reduction, improvements in soil quality and soil health, biomass products, carbon sequestration, reduced soil erosion, community aesthetic benefits, and better habitat for birds and wild animals. The potential of Miscanthus to produce sustainable feedstock for energy, or to be converted to pulp, building materials and paper is characterized.

The NATO project field site at Fort Riley, Kansas, has been used to investigate the growth of Miscanthus in soil that contains lead from past military activities (Alasmary, 2020). Miscanthus establishment was successful, and the crop grew well in the lead-contaminated soil under the climatic conditions at the site. Tilling and soil amendments were beneficial to growth and crop yield. Lead uptake into the biomass was reduced by adding biosolids as an amendment; simultaneously, soil health was improved, based on microbial numbers and composition, and organic carbon increased with time. The soil health at Fort Riley field site was investigated by assessing the effects on the nematode community of growing Miscanthus, tilling, and adding soil amendments. Significant changes were observed in trophic group structure with Miscanthus compared to the soil with mixed plant cover. Tilling the soil prior to establishment of the Miscanthus and adding biosolids as an amendment affected the nematode community and important soil processes. Tilling and tilling plus adding biosolids affected the nematode community the most. The most conserved population of the trophic structure was the bacterivores. There was also an increase in herbivores associated with the biosolids amendment (Alasmary et al., 2020).

The NATO project field site at Dolyna, Western Ukraine (former military site contaminated by various trace elements), has been used to ensure the successful cultivation of Miscanthus at the site and to investigate the impact of agricultural practices on Miscanthus biomass and phytoremediation parameters. These include incorporation of soil amendments, both mineral and organic, and/or treatment of rhizomes with Plant Growth Regulators (PGRs). Three and four years' cultivation of the crop at this military site showed good growth of Miscanthus with increasing biological parameters and harvest value (Chapter 5). Field experiments have shown the positive prospects for application of the Miscanthus phytotechnology with biomass production developed at contaminated military sites within NATO SPS MYP G4687 to other locations in Ukraine and Czech Republic where there is need to revitalize the postmilitary sites and to produce biomass for energy and bioproducts.

The NATO project research field established on marginal agricultural land at Tokarivka, Central Ukraine, illustrated the positive impact of different PGRs on Miscanthus development and harvest values on marginal land (Pidlisnyuk and Stefanovska, 2018). The successful reclamation of mined lands has been achieved by applying the best practices that are described in the book (Chapter 5).

One long-term phytotechnology option in the restoration of contaminated land is to develop it into a forest. Trees have been planted at many different sites, especially minelands, to address contamination and improve soil health. Some neglected sites eventually revert to forests if the climate is suitable. The Forest Service of the U.S. Department of Agriculture has developed the forestry reclamation approach, which has been used for successful reforestation of mined lands. A book with best management practices has been published by the U.S. Forest Service (Adams, 2017). Miscanthus has some specific advantages over woodlands when terrain permits its planting and harvest. First, it will yield useful product as quickly as even rapid-cycling poplars or willows, within 3–4 years. Because it is harvested annually, rather than on a longer rotation, it assures that biomass or income can be available every year. Second, it is far easier to remove if plans for the land are changed. This is a significant advantage on military or postmilitary lands that may be in transition. Third, Miscanthus is easily made into locally useful materials such as clean animal bedding or mulches, or used directly for combustion, and it will relatively rapidly decompose in situ without the need for chipping, unlike wood. This can more quickly and economically enhance soil tilth which is often an important component of the remediation of contaminated lands.

Economic issues are important and may limit choices at contaminated sites. Because of climate change, more decisions are being made based on sustainability and triple bottom line considerations. Adding organic carbon to the soil and improving soil health have value for society and may be included in making decisions. Since some benefits of phytoremediation such as adding soil carbon have both soil benefits and global value because of climate benefits, there is a need to include all benefits in making decisions.

Taking a wide overview of the materials in the present volume, it is becoming clear that Miscanthus production is developed to the stage that it can be treated more like a commodity than a specialty crop. Technologies are well developed for reproduction of rhizome propagules, and micropropagation has become routine. There is a good understanding of the water and weed management needs for successful establishment of fields. Weather and climate change remain uncontrollable factors but genetic technologies are available to address both cold and heat injuries to plants. While there is legitimate concern for dangers of invasiveness in seed reproduction of crops, transformation of apical meristem tissues or somatic embryos (Kim et al., 2010), followed by regeneration of a single clone, should be feasible.

An additional tool in the kit of molecular biologists is the CRISPR-Cas system, the recently discovered, Nobel prize winning, DNA editing enzyme system, which can be used to edit in or out various gene sequences. No trace of the tool remains in the product, so that it behaves as if a random spontaneous mutation. This will allow one to edit out pollen function, for instance, to convert an optimal tetraploid *M. × giganteus* into a sterile hybrid. One might also alter the flowering locus C system (Ruelens et al., 2013) or its equivalent to delay flowering time, or fully disrupt flower development in a desirable CV, so that it fails to bloom either at most latitudes of interest, or entirely. This should significantly enhance biomass yields at lower latitudes, where early flowering seems to limit biomass, as discussed in Chapter 5. We recognize that in the European context some might raise objections to these technologies, although they introduce no foreign DNA by any means other than conventional hybridization. As a strictly nonfood crop, Miscanthus is exempt from such strictures in many countries. Time and necessity may also change minds.

12.2 Recommendations

Ideally all land should be used for beneficial purposes. Improving soil health and increasing organic carbon in soil should be high priorities because of both local and global benefits. Further research and development of phytotechnologies with biomass production is recommended, including additional research with Miscanthus. An important issue is how best to use Miscanthus biomass not only for energy production but also for conversion to different bioproducts. High yields with prominent content of lignocellulose, low requirement for nutrient inputs, and low susceptibility to pests and diseases make Miscanthus an excellent feedstock for producing fiber based materials such as construction or paper industry products. Future research and practice should concentrate on looking into the full Miscanthus chain starting from production while ensuring sustainable land management, through ensuring optimal conditions for storing biomass, with development of proper technology for biomass processing into fibrous materials, pulp, and paper when Miscanthus has been harvested from marginal or contaminated soils. Some of the genetic techniques discussed above promise to enhance sustainable land management, allowing larger production of renewable energy and biomaterial without increasing land requirements. This is an essential component of a transition to a fully sustainable societal energy (and Miscanthus) cycle.

References

Adams, M. B., Ed. 2017. *The forestry reclamation approach: Guide to successful reforestation of mined lands*. Report NRS-169, Forest Service, Newtown Square, PA. https://doi.org/10.2737/NRS-GTR-169

Alasmary, Z. 2020. *Laboratory to field-scale investigations to evaluate phosphate amendments and Miscanthus for phytostabilization of lead-contaminated military sites*. Ph.D. Dissertation, Kansas State University, Manhattan.

Alasmary, Z., Todd, T., Hettiarachchi, G., et al. 2020. Effect of soil treatments and amendments on the nematode community under Miscanthus growing in a lead contaminated military site. *Agronomy*, 10, 1717. doi: 10.3390/agronomy10111727.

Kim, H. S., Zhang, G., Juvik, J. A., & Widholm, J. M. 2010. *Miscanthus x giganteus* plant regeneration: Effect of callus types, ages and culture methods on regeneration competence. *GCB Bioenergy*, 2, 192–200. doi: 10.1111/j.1757–1707.2010.01054.x.

Nobel Prize. (2020). *The Nobel Prize in Chemistry 2020*. https://www.google.com/url?sa=t&rct=j&q=&esrc=s&source=web&cd=&cad=rja&uact=8&ved=2ahUKEwin3rDlorLtAhVEPq0KHSjTCUEQFjAAegQIBRAC&url=https%3A%2F%2Fwww.nobelprize.org%2Fprizes%2Fchemistry%2F2020%2Fpress-release%2F&usg=AOvVaw28iDZV-f4W3kNm6NyW6o8V. Accessed December 3, 2020.

Pidlisnyuk, V., & Stefanovska, T. (2018). *Methods for growing M. × giganteus at the abandoned land* (Patent No. 127487), Ukraine.

Ruelens, P., de Maagd, R., Proost, S. et al. 2013. FLOWERING LOCUS C in monocots and the tandem origin of angiosperm-specific MADS-box genes. *Nature Communications*, 4, 2280. doi: 10.1038/ncomms3280.

Index